The Earliest Arithmetics in English

Early English Text Society.
Extra Series, No. CXVIII.
1922 (for 1916).

The Earliest Arithmetics in English

EDITED WITH INTRODUCTION

BY

ROBERT STEELE

LONDON:
PUBLISHED FOR THE EARLY ENGLISH TEXT SOCIETY
BY HUMPHREY MILFORD, OXFORD UNIVERSITY PRESS,
AMEN CORNER, E.C. 4.

OXFORD
UNIVERSITY PRESS

Great Clarendon Street, Oxford OX2 6DP
United Kingdom

Oxford University Press is a department of the University of Oxford.
It furthers the University's objective of excellence in research, scholarship,
and education by publishing worldwide. Oxford is a registered trade mark of
Oxford University Press in the UK and in certain other countries

© The Early English Text Society 1922 (for 1916)

The moral rights of the authors have been asserted

Database right Oxford University Press (maker)

First Edition published in 1922 (for 1916)

All rights reserved. No part of this publication may be reproduced,
stored in a retrieval system, or transmitted, in any form or by any means,
without the prior permission in writing of Oxford University Press,
or as expressly permitted by law, or under terms agreed with the appropriate
reprographics rights organization. Enquiries concerning reproduction
outside the scope of the above should be sent to the Rights Department,
Oxford University Press, at the address above

You must not circulate this book in any other form
and you must impose this same condition on any acquirer

Published in the United States of America by Oxford University Press
198 Madison Avenue, New York, NY 10016, United States of America

British Library Cataloguing in Publication Data
Data available

Library of Congress Cataloging in Publication Data
Data available

Extra Series, 118

ISBN 978-0-85-991716-2

INTRODUCTION

THE number of English arithmetics before the sixteenth century is very small. This is hardly to be wondered at, as no one requiring to use even the simplest operations of the art up to the middle of the fifteenth century was likely to be ignorant of Latin, in which language there were several treatises in a considerable number of manuscripts, as shown by the quantity of them still in existence. Until modern commerce was fairly well established, few persons required more arithmetic than addition and subtraction, and even in the thirteenth century, scientific treatises addressed to advanced students contemplated the likelihood of their not being able to do simple division. On the other hand, the study of astronomy necessitated, from its earliest days as a science, considerable skill and accuracy in computation, not only in the calculation of astronomical tables but in their use, a knowledge of which latter was fairly common from the thirteenth to the sixteenth centuries.

The arithmetics in English known to me are :—

(1) Bodl. 790 G. VII. (2653) f. 146-154 (15th c.) *inc.* "Of augrym ther be IX figures in numbray . . ." A mere unfinished fragment, only getting as far as Duplation.
(2) Camb. Univ. LI. IV. 14 (III.) f. 121-142 (15th c.) *inc.* "Al maner of thyngis that prosedeth ffro the frist begynnyng . . ."
(3) Fragmentary passages or diagrams in Sloane 213 f. 120-3 (a fourteenth-century counting board), Egerton 2852 f. 5-13, Harl. 218 f. 147 and
(4) The two MSS. here printed; Eg. 2622 f. 136 and Ashmole 396 f. 48. All of these, as the language shows, are of the fifteenth century.

The CRAFTE OF NOMBRYNGE is one of a large number of scientific treatises, mostly in Latin, bound up together as Egerton MS. 2622 in the British Museum Library. It measures 7" × 5", 29-30 lines to the page, in a rough hand. The English is N.E. Midland in dialect. It is a translation and amplification of one of the numerous glosses on the *de algorismo* of Alexander de Villa Dei (c. 1220), such as that of

Thomas of Newmarket contained in the British Museum MS. Reg. 12, E. 1. A fragment of another translation of the same gloss was printed by Halliwell in his *Rara Mathematica* (1835) p. 29.* It corresponds, as far as p. 71, l. 2, roughly to p. 3 of our version, and from thence to the end p. 2, ll. 16-40.

The ART OF NOMBRYNG is one of the treatises bound up in the Bodleian MS. Ashmole 396. It measures $11\frac{1}{2}'' \times 17\frac{3}{4}''$, and is written with thirty-three lines to the page in a fifteenth century hand. It is a translation, rather literal, with amplifications of the *de arte numerandi* attributed to John of Holywood (Sacrobosco) and the translator had obviously a poor MS. before him. The *de arte numerandi* was printed in 1488, 1490 (*s.n.*), 1501, 1503, 1510, 1517, 1521, 1522, 1523, 1582, and by Halliwell separately and in his two editions of *Rara Mathematica*, 1839 and 1841, and reprinted by Curze in 1897.

Both these tracts are here printed for the first time, but the first having been circulated in proof a number of years ago, in an endeavour to discover other manuscripts or parts of manuscripts of it, Dr. David Eugene Smith, misunderstanding the position, printed some pages in a curious transcript with four facsimiles in the *Archiv für die Geschichte der Naturwissenschaften und der Technik*, 1909, and invited the scientific world to take up the "not unpleasant task" of editing it.

ACCOMPTYNGE BY COUNTERS is reprinted from the 1543 edition of Robert Record's Arithmetic, printed by R. Wolfe. It has been reprinted within the last few years by Mr. F. P. Barnard, in his work on Casting Counters. It is the earliest English treatise we have on this variety of the Abacus (there are Latin ones of the end of the fifteenth century), but there is little doubt in my mind that this method of performing the simple operations of arithmetic is much older than any of the pen methods. At the end of the treatise there follows a note on merchants' and auditors' ways of setting down sums, and lastly, a system of digital numeration which seems of great antiquity and almost world-wide extension.

After the fragment already referred to, I print as an appendix the 'Carmen de Algorismo' of Alexander de Villa Dei in an enlarged and corrected form. It was printed for the first time by Halliwell in *Rara Mathematica*, but I have added a number of stanzas from

---

* Halliwell printed the two sides of his leaf in the wrong order. This and some obvious errors of transcription—'ferye' for 'ferthe,' 'lest' for 'left,' etc., have not been corrected in the reprint on pp. 70-71.

various manuscripts, selecting various readings on the principle that the verses were made to scan, aided by the advice of my friend Mr. Vernon Rendall, who is not responsible for the few doubtful lines I have conserved. This poem is at the base of all other treatises on the subject in medieval times, but I am unable to indicate its sources.

The Subject Matter.

Ancient and medieval writers observed a distinction between the Science and the Art of Arithmetic. The classical treatises on the subject, those of Euclid among the Greeks and Boethius among the Latins, are devoted to the Science of Arithmetic, but it is obvious that coeval with practical Astronomy the Art of Calculation must have existed and have made considerable progress. If early treatises on this art existed at all they must, almost of necessity, have been in Greek, which was the language of science for the Romans as long as Latin civilisation existed. But in their absence it is safe to say that no involved operations were or could have been carried out by means of the alphabetic notation of the Greeks and Romans. Specimen sums have indeed been constructed by moderns which show its possibility, but it is absurd to think that men of science, acquainted with Egyptian methods and in possession of the abacus,* were unable to devise methods for its use.

The Pre-Medieval Instruments Used in Calculation.

The following are known :—

(1) A flat polished surface or tablets, strewn with sand, on which figures were inscribed with a stylus.

(2) A polished tablet divided longitudinally into nine columns (or more) grouped in threes, with which counters were used, either plain or marked with signs denoting the nine numerals, etc.

(3) Tablets or boxes containing nine grooves or wires, in or on which ran beads.

(4) Tablets on which nine (or more) horizontal lines were marked, each third being marked off.

The only Greek counting board we have is of the fourth class and was discovered at Salamis. It was engraved on a block of marble, and measures 5 feet by $2\frac{1}{2}$. Its chief part consists of eleven parallel lines, the 3rd, 6th, and 9th being marked with a cross. Another section consists of five parallel lines, and there are three

* For Egyptian use see Herodotus, ii. 36, Plato, *de Legibus*, VII.

rows of arithmetical symbols. This board could only have been used with counters (*calculi*), preferably unmarked, as in our treatise of *Accomptynge by Counters.*

CLASSICAL ROMAN METHODS OF CALCULATION.

We have proof of two methods of calculation in ancient Rome, one by the first method, in which the surface of sand was divided into columns by a stylus or the hand. Counters (*calculi*, or *lapilli*), which were kept in boxes (*loculi*), were used in calculation, as we learn from Horace's schoolboys (Sat. 1. vi. 74). For the sand see Persius I. 131, "Nec qui abaco numeros et secto in pulvere metas scit risisse," Apul. Apolog. 16 (pulvisculo), Mart. Capella, lib. vii. 3, 4, etc. Cicero says of an expert calculator "eruditum attigisse pulverem," (de nat. Deorum, ii. 18). Tertullian calls a teacher of arithmetic "primus numerorum arenarius" (de Pallio, *in fine*). The counters were made of various materials, ivory principally, "Adeo nulla uncia nobis est eboris, etc." (Juv. XI. 131), sometimes of precious metals, "Pro calculis albis et nigris aureos argenteosque habebat denarios" (Pet. Arb. Satyricon, 33).

There are, however, still in existence four Roman counting boards of a kind which does not appear to come into literature. A typical one is of the third class. It consists of a number of transverse wires, broken at the middle. On the left hand portion four beads are strung, on the right one (or two). The left hand beads signify units, the right hand one five units. Thus any number up to nine can be represented. This instrument is in all essentials the same as the Swanpan or Abacus in use throughout the Far East. The Russian stchota in use throughout Eastern Europe is simpler still. The method of using this system is exactly the same as that of *Accomptynge by Counters*, the right-hand five bead replacing the counter between the lines.

THE BOETHIAN ABACUS.

Between classical times and the tenth century we have little or no guidance as to the art of calculation. Boethius (fifth century), at the end of lib. II. of his *Geometria* gives us a figure of an abacus of the second class with a set of counters arranged within it. It has, however, been contended with great probability that the whole passage is a tenth century interpolation. As no rules are given for its use, the chief value of the figure is that it gives the signs of the nine numbers, known as the Boethian "apices" or "notae" (from whence our word "notation"). To these we shall return later on.

THE ABACISTS.

It would seem probable that writers on the calendar like Bede (A.D. 721) and Helpericus (A.D. 903) were able to perform simple calculations; though we are unable to guess their methods, and for the most part they were dependent on tables taken from Greek sources. We have no early medieval treatises on arithmetic, till towards the end of the tenth century we find a revival of the study of science, centring for us round the name of Gerbert, who became Pope as Sylvester II. in 999. His treatise on the use of the Abacus was written (c. 980) to a friend Constantine, and was first printed among the works of Bede in the Basle (1563) edition of his works, I. 159, in a somewhat enlarged form. Another tenth century treatise is that of Abbo of Fleury (c. 988), preserved in several manuscripts. Very few treatises on the use of the Abacus can be certainly ascribed to the eleventh century, but from the beginning of the twelfth century their numbers increase rapidly, to judge by those that have been preserved.

The Abacists used a permanent board usually divided into twelve columns; the columns were grouped in threes, each column being called an "arcus," and the value of a figure in it represented a tenth of what it would have in the column to the left, as in our arithmetic of position. With this board counters or jetons were used, either plain or, more probably, marked with numerical signs, which with the early Abacists were the "apices," though counters from classical times were sometimes marked on one side with the digital signs, on the other with Roman numerals. Two ivory discs of this kind from the Hamilton collection may be seen at the British Museum. Gerbert is said by Richer to have made for the purpose of computation a thousand counters of horn; the usual number of a set of counters in the sixteenth and seventeenth centuries was a hundred.

Treatises on the Abacus usually consist of chapters on Numeration explaining the notation, and on the rules for Multiplication and Division. Addition, as far as it required any rules, came naturally under Multiplication, while Subtraction was involved in the process of Division. These rules were all that were needed in Western Europe in centuries when commerce hardly existed, and astronomy was unpractised, and even they were only required in the preparation of the calendar and the assignments of the royal exchequer. In England, for example, when the hide developed from the normal holding of a household into the unit of taxation, the calculation of the geldage in each shire required a sum in division; as we know from the fact that one of the Abacists proposes the sum: "If 200 marks are levied on the county of Essex, which contains according to Hugh of Bocland 2500 hides, how much does each hide pay?" * Exchequer methods up to the sixteenth century were founded on the abacus, though when we have details later on, a different and simpler form was used.

The great difficulty of the early Abacists, owing to the absence of a figure representing zero, was to place their results and operations in the proper columns of the abacus, especially when doing a division sum. The chief differences noticeable in their works are in the methods for this rule. Division was either done directly or by means of differences between the divisor and the next higher multiple of ten to the divisor. Later Abacists made a distinction between "iron" and "golden" methods of division. The following are examples taken from a twelfth century treatise. In following the operations it must be remembered that a figure asterisked represents a counter taken from the board. A zero is obviously not needed, and the result may be written down in words.

(a) MULTIPLICATION. 4600 × 23.

Thousands						
Hundreds	Tens	Units	Hundreds	Tens	Units	
		4	6			Multiplicand.
		1	8			600 × 3.
	1	2				4000 × 3.
	1	2				600 × 20.
	8					4000 × 20.
1		5	8			Total product.
				2	3	Multiplier.

* See on this Dr. Poole, *The Exchequer in the Twelfth Century*, Chap. III., and Haskins, *Eng. Hist. Review*, 27, 101. The hidage of Essex in 1130 was 2364 hides.

Introduction

(b) DIVISION: DIRECT. 100,000 ÷ 20,023. Here each counter in turn is a separate divisor.

Thousands			Units			
H.	T.	U.	H.	T.	U.	
	2			2	3	**Divisors.**
	2					Place greatest divisor to right of dividend.
1						**Dividend.**
	2					Remainder.
			1			
	1	9	9			Another form of same.
				8		Product of 1st Quotient and 20.
	1	9	9	2		Remainder.
				1	2	Product of 1st Quotient and 3.
	1	9	9		8	**Final remainder.**
					4	Quotient.

(c) DIVISION BY DIFFERENCES. 900 ÷ 8. Here we divide by (10−2).

H.	T.	U.	
		2	Difference.
		8	Divisor.
*9			**Dividend.**
*1	8		Product of difference by 1st Quotient (9).
		2	Product of difference by 2nd Quotient (1).
*1			Sum of 8 and 2.
	2		Product of difference by 3rd Quotient (1).
		4	Product of difference by 4th Quot. (2). **Remainder.**
		2	4th Quotient.
	1		3rd Quotient.
	1		2nd Quotient.
	9		1st Quotient.
1	1	2	**Quotient.** (Total of all four.)

* These figures are removed at the next step.

DIVISION. 7800 ÷ 166.

Thousands						
H.	T.	U.	H.	T.	U.	
				3	4	Differences (making 200 trial divisor).
			1	6	6	Divisors.
		*7	8			**Dividends.**
		1				Remainder of greatest dividend.
				1	2	Product of 1st difference (4) by 1st Quotient (3).
			9			Product of 2nd difference (3) by 1st Quotient (3).
		*2	8	2		New dividends.
				3	4	Product of 1st and 2nd difference by 2nd Quotient (1).
		*1	1	6		New dividends.
				2		Product of 1st difference by 3rd Quotient (5).
				1	5	Product of 2nd difference by 3rd Quotient (5).
			*3	3		New dividends.
			1			Remainder of greatest dividend.
				3	4	Product of 1st and 2nd difference by 4th Quotient (1).
			1	6	4	**Remainder** (less than divisor).
					1	4th Quotient.
					5	3rd Quotient.
				1		2nd Quotient.
				3		1st Quotient.
				4	6	**Quotient.**

* These figures are removed at the next step.

DIVISION. 8000 ÷ 606.

Thousands						
H.	T.	U.	H.	T.	U.	
				9		Difference (making 700 trial divisor).
					4	Difference.
			6		6	Divisors.
		*8				**Dividend.**
		1				Remainder of dividend.
			9	4		Product of difference 1 and 2 with 1st Quotient (1).
		*1	9	4		New dividends.
			3			Remainder of greatest dividend.
				9	4	Product of difference 1 and 2 with 2nd Quotient (1).
		*1	3	3	4	New dividends.
			3			Remainder of greatest dividend.
				9	4	Product of difference 1 and 2 with 3rd Quotient (1).
			7	2	8	New dividends.
			6		6	Product of divisors by 4th Quotient (1).
			1	2	2	**Remainder.**
					1	4th Quotient.
					1	3rd Quotient.
					1	2nd Quotient.
				1		1st Quotient.
				1	3	**Quotient.**

* These figures are removed at the next step.

The chief Abacists are Gerbert (tenth century), Abbo, and Hermannus Contractus (1054), who are credited with the revival of the art, Bernelinus, Gerland, and Radulphus of Laon (twelfth century). We know as English Abacists, Robert, bishop of Hereford, 1095, "abacum et lunarem compotum et celestium cursum astrorum rimatus," Turchillus Compotista (Thurkil), and through him of Guilielmus R. . . . "the best of living computers," Gislebert, and Simonus de Rotellis (Simon of the Rolls). They flourished most probably in the first quarter of the twelfth century, as Thurkil's treatise deals also with fractions. Walcher of Durham, Thomas of York, and Samson of Worcester are also known as Abacists.

Finally, the term Abacists came to be applied to computers by manual arithmetic. A MS. Algorithm of the thirteenth century (Sl. 3281, f. 6, b), contains the following passage: "Est et alius modus secundum operatores sive practicos, quorum unus appellatur Abacus; et modus ejus est in computando per digitos et junctura manuum, et iste utitur ultra Alpes."

In a composite treatise containing tracts written A.D. 1157 and 1208, on the calendar, the abacus, the manual calendar and the manual abacus, we have a number of the methods preserved. As an example we give the rule for multiplication (Claud. A. IV., f. 54 vo). "Si numerus multiplicat alium numerum auferatur differentia majoris a minore, et per residuum multiplicetur articulus, et una differentia per aliam, et summa proveniet." Example, 8×7. The difference of 8 is 2, of 7 is 3, the next article being 10; 7–2 is 5. $5 \times 10 = 50$; $2 \times 3 = 6$. $50 + 6 = 56$ answer. The rule will hold in such cases as 17×15 where the article next higher is the same for both, *i.e.*, 20; but in such a case as 17×9 the difference for each number must be taken from the higher article, *i.e.*, the difference of 9 will be 11.

THE ALGORISTS.

Algorism (augrim, augrym, algram, agram, algorithm), owes its name to the accident that the first arithmetical treatise translated from the Arabic happened to be one written by Al-Khowarazmi in the early ninth century, "de numeris Indorum," beginning in its Latin form "Dixit Algorismi. . . ." The translation, of which only one MS. is known, was made about 1120 by Adelard of Bath, who also wrote on the Abacus and translated with a commentary Euclid from the Arabic. It is probable that another version was made by Gerard of Cremona (1114–1187); the number of important works that were not translated more than once from the Arabic decreases every year with our knowledge of medieval texts. A few lines of this translation, as copied by Halliwell, are given on p. 72, note 2. Another translation still seems to have been made by Johannes Hispalensis.

Algorism is distinguished from Abacist computation by recognising seven rules, Addition, Subtraction, Duplation, Mediation, Multiplication, Division, and Extraction of Roots, to which were afterwards

Introduction

added Numeration and Progression. It is further distinguished by the use of the zero, which enabled the computer to dispense with the columns of the Abacus. It obviously employs a board with fine sand or wax, and later, as a substitute, paper or parchment; slate and pencil were also used in the fourteenth century, how much earlier is unknown.* Algorism quickly ousted the Abacus methods for all intricate calculations, being simpler and more easily checked: in fact, the astronomical revival of the twelfth and thirteenth centuries would have been impossible without its aid.

The number of Latin Algorisms still in manuscript is comparatively large, but we are here only concerned with two—an Algorism in prose attributed to Sacrobosco (John of Holywood) in the colophon of a Paris manuscript, though this attribution is no longer regarded as conclusive, and another in verse, most probably by Alexander de Villedieu (Villa Dei). Alexander, who died in 1240, was teaching in Paris in 1209. His verse treatise on the Calendar is dated 1200, and it is to that period that his Algorism may be attributed; Sacrobosco died in 1256 and quotes the verse Algorism. Several commentaries on Alexander's verse treatise were composed, from one of which our first tractate was translated, and the text itself was from time to time enlarged, sections on proofs and on mental arithmetic being added. We have no indication of the source on which Alexander drew; it was most likely one of the translations of Al-Khowarasmi, but he has also the Abacists in mind, as shewn by preserving the use of differences in multiplication. His treatise, first printed by Halliwell-Phillipps in his *Rara Mathematica*, is adapted for use on a board covered with sand, a method almost universal in the thirteenth century, as some passages in the algorism of that period already quoted show: "Est et alius modus qui utitur apud Indos, et doctor hujusmodi ipsos erat quidem nomine Algus. Et modus suus erat in computando per quasdam figuras scribendo in pulvere. . . ." "Si voluerimus depingere in pulvere predictos digitos secundum consuetudinem algorismi . . ." "et sciendum est quod in nullo loco minutorum sive secundorum . . . in pulvere debent scribi plusquam sexaginta."

Modern Arithmetic.

Modern Arithmetic begins with Leonardi Fibonacci's treatise "de Abaco," written in 1202 and re-written in 1228. It is modern

---

* Slates are mentioned by Chaucer, and soon after (1410) Prosdocimo de Beldamandi speaks of the use of a "lapis" for making notes on by calculators.

rather in the range of its problems and the methods of attack than in mere methods of calculation, which are of its period. Its sole interest as regards the present work is that Leonardi makes use of the digital signs described in Record's treatise on *The arte of nombrynge by the hand* in mental arithmetic, calling it "modus Indorum." Leonardo also introduces the method of proof by "casting out the nines."

DIGITAL ARITHMETIC.

The method of indicating numbers by means of the fingers is of considerable age. The British Museum possesses two ivory counters marked on one side by carelessly scratched Roman numerals IIIV and VIIII, and on the other by carefully engraved digital signs for 8 and 9. Sixteen seems to have been the number of a complete set. These counters were either used in games or for the counting board, and the Museum ones, coming from the Hamilton collection, are undoubtedly not later than the first century. Frohner has published in the *Zeitschrift des Münchener Alterthumsvereins* a set, almost complete, of them with a Byzantine treatise; a Latin treatise is printed among Bede's works. The use of this method is universal through the East, and a variety of it is found among many of the native races in Africa. In medieval Europe it was almost restricted to Italy and the Mediterranean basin, and in the treatise already quoted (Sloane 3281) it is even called the Abacus, perhaps a memory of Fibonacci's work.

Methods of calculation by means of these signs undoubtedly have existed, but they were too involved and liable to error to be much used.

THE USE OF "ARABIC" FIGURES.

It may now be regarded as proved by Bubnov that our present numerals are derived from Greek sources through the so-called Boethian "apices," which are first found in late tenth century manuscripts. That they were not derived directly from the Arabic seems certain from the different shapes of some of the numerals, especially the 0, which stands for 5 in Arabic. Another Greek form existed, which was introduced into Europe by John of Basingstoke in the thirteenth century, and is figured by Matthew Paris (V. 285); but this form had no success. The date of the introduction of the zero has been hotly debated, but it seems obvious that the twelfth century Latin translators from the Arabic were

Introduction

perfectly well acquainted with the system they met in their Arabic text, while the earliest astronomical tables of the thirteenth century I have seen use numbers of European and not Arabic origin. The fact that Latin writers had a convenient way of writing hundreds and thousands without any cyphers probably delayed the general use of the Arabic notation. Dr. Hill has published a very complete survey of the various forms of numerals in Europe. They began to be common at the middle of the thirteenth century and a very interesting set of family notes concerning births in a British Museum manuscript, Harl. 4350 shows their extension. The first is dated $\overset{c}{\text{Mij}}$. lviii., the second $\overset{c}{\text{Mij}}$. lxi., the third $\overset{c}{\text{Mij}}$. 63, the fourth 1264, and the fifth 1266. Another example is given in a set of astronomical tables for 1269 in a manuscript of Roger Bacon's works, where the scribe began to write MCC6. and crossed out the figures, substituting the "Arabic" form.

THE COUNTING BOARD.

The treatise on pp. 52–65 is the only one in English known on the subject. It describes a method of calculation which, with slight modifications, is current in Russia, China, and Japan, to-day, though it went out of use in Western Europe by the seventeenth century. In Germany the method is called "Algorithmus Linealis," and there are several editions of a tract under this name (with a diagram of the counting board), printed at Leipsic at the end of the fifteenth century and the beginning of the sixteenth. They give the nine rules, but "Capitulum de radicum extractione ad algoritmum integrorum reservato, cujus species per ciffrales figuras ostenduntur ubi ad plenum de hac tractabitur." The invention of the art is there attributed to Appulegius the philosopher.

The advantage of the counting board, whether permanent or constructed by chalking parallel lines on a table, as shown in some sixteenth-century woodcuts, is that only five counters are needed to indicate the number nine, counters on the lines representing units, and those in the spaces above representing five times those on the line below. The Russian abacus, the "tchatui" or "stchota" has ten beads on the line; the Chinese and Japanese "Swanpan" economises by dividing the line into two parts, the beads on one side representing five times the value of those on the other. The "Swanpan" has usually many more lines than the "stchota," allowing for more extended calculations, see Tylor, *Anthropology* (1892), p. 314.

Record's treatise also mentions another method of counter notation (p. 64) "merchants' casting" and "auditors' casting." These were adapted for the usual English method of reckoning numbers up to 200 by scores. This method seems to have been used in the Exchequer. A counting board for merchants' use is printed by Halliwell in *Rara Mathematica* (p. 72) from Sloane MS. 213, and two others are figured in Egerton 2622 f. 82 and f. 83. The latter is said to be "novus modus computandi secundum inventionem Magistri Thome Thorleby," and is in principle, the same as the "Swanpan."

The Exchequer table is described in the *Dialogus de Scaccario* (Oxford, 1902), p. 38.

The Earliest Arithmetics in English.

The Crafte of Nombrynge.

Egerton 2622.

[1] **Hec algorismus ars presens dicitur; in qua   [1] leaf 136 a.
Talibus indorum fruimur bis quinque figuris.**

This boke is called þe boke of algorym, or Augrym after lewder A derivation of Algorism.
4 vse. And þis boke tretys þe Craft of Nombryng, þe quych crafte
is called also Algorym. Ther was a kyng of Inde, þe quich heyth
Algor, & he made þis craft. And after his name he called hit
algorym; or els anoþer cause is quy it is called Algorym, for þe
8 latyn word of hit s. Algorismus comes of Algos, grece, quid est Another derivation of the word.
ars, latine, craft on englis, and rides, quid est numerus, latine, A
nombur on englys, inde dicitur Algorismus per addicionem huius
sillabe mus & subtraccionem d & e, quasi ars numerandi. ¶ fforther-
12 more ȝe most vndirstonde þat in þis craft ben vsid teen figurys,
as here bene writen for ensampul, φ 9 8 7 6 5 4 3 2 1. ¶ Expone
þe too versus afore: this present craft ys called Algorismus, in þe
quych we vse teen signys of Inde. Questio. ¶ Why ten fyguris
16 of Inde? Solucio. for as I haue sayd afore þai were fonde fyrst
in Inde of a kynge of þat Cuntre, þat was called Algor.

¶ **Prima significat unum; duo vero secunda:**   versus [in margin].
¶ **Tercia significat tria; sic procede sinistre.**
20 ¶ **Donec ad extremam venias, que cifra vocatur.**

¶ **Capitulum primum de significacione figurarum.**   Expositio versus.

In þis verse is notifide þe significacion of þese figuris. And þus
expone the verse. þe first signifiyth one, þe secunde signi[2]fiyth   [2] leaf 136 b.
24 tweyne, þe thryd signifiyth thre, & the fourte signifiyth 4. ¶ And   The meaning and place of the figures.
so forthe towarde þe lyft syde of þe tabul or of þe boke þat þe
figures bene writen in, til þat þou come to the last figure, þat is

Notation and Numeration.

called a cifre. ¶ Questio. In quych syde sittes þe first figure? Solucio, forsothe loke quich figure is first in þe ryȝt side of þe bok or of þe tabul, & þat same is þe first figure, for þou schal write bakeward, as here, 3. 2. 6. 4. 1. 2. 5. The figure of 5. was first write, & he is þe first, for he sittes oꞃ þe riȝt syde. And the figure of 3 is last. ¶ Neuer-þe-les wen he says ¶ Prima significat vnum &c., þat is to say, þe first betokenes one, þe secunde. 2. & fore-þer-more, he vndirstondes noȝt of þe first figure of euery rew. ¶ But he vndirstondes þe first figure þat is in þe nombur of þe forsayd teen figuris, þe quych is one of þese. 1. And þe secunde 2. & so forth.

Which figure is read first.

4

8

versus [in margin].

¶ **Quelibet illarum si primo limite ponas,**
 ¶ **Simpliciter se significat: si vero secundo,**
  Se decies: sursum procedas multiplicando.
 ¶ **Namque figura sequens quamuis signat decies plus.**
 ¶ **Ipsa locata loco quam significat pertinente.**

12

16

Expositio [in margin].

An explanation of the principles of notation.

¹ leaf 137 a.

An example: units,

tens,

hundreds,

thousands.

¶ Expone þis verse þus. Euery of þese figuris bitokens hym selfe & no more, yf he stonde in þe first place of þe rewele / this worde Simpliciter in þat verse it is no more to say but þat, & no more. ¶ If it stonde in the secunde place of þe rewle, he betokens tene tymes hym selfe, as þis figure 2 here 20 tokens ten tyme hym selfe, ¹þat is twenty, for he hym selfe betokenes tweyne, & ten tymes twene is twenty. And for he stondis oꞃ þe lyft side & in þe secunde place, he betokens ten tyme hym selfe. And so go forth. ¶ ffor euery figure, & he stonde aftur a-noþer toward the lyft side, he schal betokene ten tymes as mich more as he schul betoken & he stode in þe place þere þat þe figure a-fore hym stondes. loo an ensampulle. 9. 6. 3. 4. þe figure of 4. þat hase þis schape 4. betokens bot hymselfe, for he stondes in þe first place. The figure of 3. þat hase þis schape 3. betokens ten tymes more þen he schuld & he stode þere þat þe figure of 4. stondes, þat is thretty. The figure of 6, þat hase þis schape 6, betokens ten tymes more þan he schuld & he stode þere as þe figure of 3. stondes, for þere he schuld tokyne bot sexty, & now he betokens ten tymes more, þat is sex hundryth. The figure of 9. þat hase þis schape 9. betokens ten tymes more þane he schuld & he stode in þe place þere þe figure of sex stondes, for þen he schuld betoken to 9. hundryth, and in þe place þere he stondes now he betokens 9. þousande. Al þe hole nombur is 9 thousande sex hundryth & foure & thretty. ¶ ffortherrmore, when

20

24

28

32

36

40

The Three Kinds of Numbers.

þou schalt rede a nombur of figure, þou schalt begyne at þe last figure in the lyft side, & rede so forth to þe riȝt side as here 9. 5. 3. 4. Thou schal begyn to rede at þe figure of 9. & rede forth 4 þus. 9. ¹thousand sex hundryth thritty & foure. But when þou schalle write, þou schalt be-gynne to write at þe ryȝt side. *How to read the number.* ¹ leaf 137 b.

¶ **Nil cifra significat sed dat signare sequenti.**

Expone þis verse. A cifre tokens noȝt, bot he makes þe figure to betoken þat comes aftur hym more þan he schuld & he were away, as þus 1ɸ. here þe figure of one tokens ten, & yf þe cifre were away² & no figure by-fore hym he schuld token bot one, for þan he schuld stonde in þe first place. ¶ And þe cifre tokens nothyng hym selfe. for al þe nombur of þe ylke too figures is bot ten. ¶ Questio. Why says he þat a cifre makys a figure to signifye (tyf) more &c. ¶ I speke for þis worde significatyf, ffor sothe it may happe aftur a cifre schuld come a-noþur cifre, as þus 2ɸɸ. And ȝet þe secunde cifre shuld token neuer þe more excep he schuld kepe þe order of þe place. and a cifre is no figure significatyf. *The meaning and use of the cipher.*

¶ **Quam precedentes plus ultima significabit /**

Expone þis verse þus. þe last figure schal token more þan alle þe oþer afore, thouȝt þere were a hundryth thousant figures afore, as þus, 16798. þe last figure þat is 1. betokens ten thousant. And alle þe oþer figures ben bot betokene bot sex thousant seuyne hundryth nynty & 8. ¶ And ten thousant is more þen alle þat nombur, ergo þe last figure tokens more þan all þe nombur afore. *The last figure means more than all the others, since it is of the highest value.*

³¶ **Post predicta scias breuiter quod tres numerorum Distincte species sunt; nam quidam digiti sunt; Articuli quidam; quidam quoque compositi sunt.** ³ leaf 138 a.

¶ **Capitulum 2ᵐ de triplice divisione numerorum.**

¶ The auctor of þis tretis departys þis worde a nombur into 3 partes. Some nombur is called digitus latine, a digit in englys. Somme nombur is called articulus latine. An Articul in englys. Some nombur is called a composyt in englys. ¶ Expone þis verse. know þou aftur þe forsayd rewles þat I sayd afore, þat þere ben thre spices of nombur. Oone is a digit, Anoþer is an Articul, & þe toþer a Composyt. versus. *Digits. Articles. Composites.*

¶ **Sunt digiti numeri qui citra denarium sunt.**

¶ Here he telles qwat is a digit, Expone versus sic. Nomburs digitus bene alle nomburs þat ben with-inne ten, as nyne, 8. 7. 6. 5. 4. 3. 2. 1. *What are digits.*

² In MS. 'awiy.'

Digits, Articles, and Composites.

¶ **Articupli decupli degito**rum**; compositi s**u**nt
Illi qui constant ex articulis degitisq**ue.

¶ Here he telles what is a composyt and what is an*e* articul.

What are articles.
Expone sic v*er*sus. ¶ Articulis ben[1] alle þat may be deuidyt into nomb*urs* of ten & nothynge leue ouer, as twenty, thretty, fourty,
a hundryth, a thousand, & such oþ*er*, ffor twenty may be depa*r*tyt
in-to 2 nomb*urs* of ten, fforty in to foure nomb*urs* of ten, & so forth.

[2] *leaf 138 b.*
What numbers are composites.
[2]Compositys ben) nomb*urs* þat bene componyt of a digyt & of an
articulle as fouretene, fyftene, sextene, & such oþ*er*. ffortene is
componyd of foure þat is a digit & of ten þat is an articulle.
ffiftene is componyd of 5 & ten, & so of all oþ*er*, what þat þai ben.
Short-lych eue*ry* nomb*ur* þat be-gynnes wiþ a digit & endyth in a
articulle is a composyt, as fortene bygennynge by foure þat is a
digit, & endes in ten.

¶ **Ergo, p**roposito nu*mer*o **tibi scribere, p**rimo
Respicias quid sit numerus**; si digitus sit**
P*r*imo **scribe loco digitu**m**, si compositus sit**
P*r*imo **scribe loco digitu**m **post articulu**m**; sic.**

How to write a number,
¶ here he telles how þou schalt wyrch whan þou schalt write a
nomb*ur*. Expone v*er*sum sic, & fac iuxta exponen*tis* sentencia*m*;
whan þou hast a nomb*ur* to write, loke fyrst what maner nomb*ur*
it ys þ*at* þou schalt write, whether it be a digit or a composit or an
if it is a digit;
Articul. ¶ If he be a digit, write a digit, as yf it be seuen, write
seuen & write þat digit in þe first place toward þe ryght side. If it
if it is a composite.
be a composyt, write þe digit of þe composit in þe first place &
write þe articul of þat digit in þe secunde place next toward þe lyft
side. As yf þou schal write sex & twenty. write þe digit of þe
nomb*ur* in þe first place þat is sex, and write þe articul next aft*ur*
þat is twenty, as þus 26. But whan þou schalt sowne or speke
[3] *leaf 139 a.*
How to read it.
[3]or rede an Composyt þou schalt first sowne þe articul & aft*ur* þe
digit, as þou seyst by þe comyne speche, Sex & twenty & nouȝt
twenty & sex. v*er*sus.

¶ **Articul**us **si sit, in p**rimo **limite cifram,
Articulu**m vero **reliqu**is **inscr**i**be figur**is.

How to write Articles:
¶ Here he tells how þou schal write when þe nombre þat þou
hase to write is an Articul. Expone v*er*sus sic & fac secundu*m*
sentenciam. Ife þe nomb*ur* þat þou hast write be an Articul, write
tens,
first a cifre & aft*ur* þe cifer write an Articulle þus. 2φ. fforther-
more þou schalt vndi*r*stonde yf þou haue an Articul, loke how

---

[1] 'ben' repeated in MS.

4

8

12

16

20

24

28

32

36

The Seven Rules of Arithmetic.

mych he is, yf he be with-ynne an hundryth, þou schalt write bot
one cifre, afore, as here .9φ. If þe articulle be by hym-silfe & be hundreds,
an hundrid euene, þen schal þou write .1. & 2 cifers afore, þat he
may stonde in þe thryd place, for euery figure in þe thryd place
schal token a hundrid tymes hym selfe. If þe articul be a thousant thousands,
or thousandes[1] and he stonde by hym selfe, write afore 3 cifers & so &c.
forþ of al oþer.

¶ **Quolibet in numero, si par sit prima figura,**
 Par erit & totum, quicquid sibi continuatur;
 Impar si fuerit, totum tunc fiet et impar.

¶ Here he teches a generalle rewle þat yf þe first figure in þe To tell an
rewle of figures token a nombur þat is euene al þat nombur of even number
figurys in þat rewle schal be euene, as here þou may see 6. 7. 3. 5. 4.
Computa & proba. ¶ If þe first [2]figure token an nombur þat is ode, ² leaf 139 b.
alle þat nombur in þat rewle schalle be ode, as here 5 6 7 8 6 7. or an odd.
Computa & proba. versus.

¶ **Septem sunt partes, non plures, istius artis;**
¶ **Addere, subtrahere, duplare, dimidiare,**
 Sextaque diuidere, sed quinta multiplicare;
 Radicem extrahere pars septima dicitur esse.

¶ Here telles þat þer ben .7. spices or partes of þis craft. The The seven
first is called addicion, þe secunde is called subtraccion. The thryd rules.
is called duplacion. The 4. is called dimydicion. The 5. is called
multiplicacion. The 6 is called diuision. The 7. is called extraccion
of þe Rote. What all þese spices bene hit schalle be tolde singillatim in here caputule.

¶ **Subtrahis aut addis a dextris vel mediabis:**

Thou schal be-gynne in þe ryght side of þe boke or of a tabul. Add, sub-
loke were þou wul be-gynne to write latyn or englys in a boke, & tract, or
þat schalle be called þe lyft side of the boke, þat þou writest toward halve, from
þat side schal be called þe ryght side of þe boke. Versus. right to lett.

A leua dupla, diuide, multiplica.

Here he telles þe in quych side of þe boke or of þe tabul þou
schalle be-gyne to wyrch duplacion, diuision, and multiplicacion.
Thou schal begyne to worch in þe lyft side of þe boke or of þe Multiply or
tabul, but yn what wyse þou schal wyrch in hym dicetur singil- divide from
latim in sequentibus capitulis et de vtilitate cuiuslibet artis & left to right.
sic Completur [3]prohemium & sequitur tractatus & primo de arte ³ leaf 140.
addicionis que prima ars est in ordine.

---

[1] In MS. 'thausandes.'

The Craft of Addition.

Addere si numero numerum vis, ordine tali
Incipe; scribe duas primo series numerorum
Primam sub prima recte ponendo figuram,
Et sic de reliquis facias, si sint tibi plures.

Four things must be known:
¶ Here by-gynnes þe craft of Addicioñ. In þis craft þou most knowe foure thynges. ¶ Fyrst þou most know what is addicioñ. Next þou most know how mony rewles of figurys þou most haue. ¶ Next þou most know how mony diuers casys happes in þis craft of addicioñ. ¶ And next qwat is þe profet of þis craft. ¶ As for

what it is;
þe first þou most know þat addicioñ is a castyng to-gedur of twoo nomburys in-to one nombre. As yf I aske qwat is twene & thre. þou wyl cast þese twene nombres to-gedur & say þat it is fyue.

how many rows of figures;
¶ As for þe secunde þou most know þat þou schalle haue tweyne rewes of figures, one vndur a-nother, as here þou mayst se. 1234

how many cases;
¶ As for þe thryd þou most know þat there ben foure diuerse 2168. cases. As for þe forthe þou most know þat þe profet of þis craft is

what is its result.
to telle what is þe hole nombur þat comes of diuerse nomburis. Now as to þe texte of oure verse, he teches there how þou schal worch in þis craft. ¶ He says yf þou wilt cast one nombur to anoþer nombur, þou most by-gynne on þis wyse. ¶ ffyrst write

[1] *leaf 140 b.*
How to set down the sum.
[1]two rewes of figuris & nombris so þat þou write þe first figure of þe hyer nombur euene vndir the first figure of þe nether nombur, And þe secunde of þe nether nombur euene vndir þe secunde of þe hyer, & so forthe of euery figure of both þe rewes as þou mayst se   123
                                                                                                           234.

¶ Inde duas adde primas hac condicione:
Si digitus crescat ex addicione priorum;
Primo scribe loco digitum, quicunque sit ille.

Add the first figures;
¶ Here he teches what þou schalt do when þou hast write too rawes of figuris on vnder an-oþer, as I sayd be-fore. ¶ He says þou schalt take þe first figure of þe heyer nombre & þe fyrst figure of þe neþer nombre, & cast hem to-geder vp-on þis condicion. Thou schal loke qweþer þe nomber þat comys þere-of be a digit or no.

rub out the top figure;
¶ If he be a digit þou schalt do away þe first figure of þe hyer nombre, and write þere in his stede þat he stode Inne þe digit, þat

write the result in its place.
comes of þe ylke 2 figures, & so wrich forth on oþer figures yf þere be ony moo, til þou come to þe ende toward þe lyft side. And lede þe nether figure stonde still euer-more til þou haue ydo. ffor þere-by þou schal wyte wheþer þou hast done wel or no, as I schal tell þe afterward in þe ende of þis Chapter. ¶ And loke allgate

[2] *leaf 141 a.*
þat þou be-gynne to worch in þis Craft of Addi[2]cion in þe ry3t side, 40

4

8

12

16

20

24

28

32

36

The Cases of the Craft of Addition.

here is an ensampul of þis case $\frac{1234}{2142}$. Caste 2 to foure & þat wel be *Here is an example.*
sex, do away 4. & write in þe same place þe figure of sex.
¶ And lete þe figure of 2 in þe nether rewe stonde stil. When
þou hast do so, cast 3 & 4 to-geder and þat wel be seuen þat is
a digit. Do away þe 3, & set þere seuen, and lete þe neþer figure
stonde stille, & so worch forth bakward til þou hast ydo all to-geder.

**Et si compositus, in limite scribe sequente
Articulum, primo digitum; quia sic iubet ordo.**

¶ Here is þe secunde case þat may happe in þis craft. And þe
case is þis, yf of þe casting of 2 nomburis to-geder, as of þe figure of *Suppose it is a Composite,*
þe hyer rewe & of þe figure of þe neþer rewe come a Composyt, how *set down the digit,*
schalt þou worch. þus þou schalt worch. Thou shalt do away þe *and carry the tens.*
figure of þe hyer nomber þat was cast to þe figure of þe neþer
nomber. ¶ And write þere þe digit of þe Composyt. And set þe
articul of þe composit next after þe digit in þe same rewe, yf þere
be no mo figures after. But yf þere be mo figuris after þat digit.
And þere he schall be rekend for hym selfe. And when þou schalt
adde þat ylke figure þat berys þe articulle ouer his hed to þe figure
vnder hym, þou schalt cast þat articul to þe figure þat hase hym ouer
his hed, & þere þat Articul schal token hym selfe. Io an Ensam- *Here is an example.*
pull [1] of all 326. Cast 6 to 6, & þere of wil arise twelue. do away *¹ leaf 141 b.*
þe hyer 6 $\frac{216}{}$ & write þere 2, þat is þe digit of þis composit.
And þen write þe articulle þat is ten ouer þe figuris hed of twene
as þus $\frac{1}{322}$ Now cast þe articulle þat standus vpon þe figuris of
twene 216. hed to þe same figure, & reken þat articul bot for one,
and þan þere wil arise thre. þan cast þat thre to þe neþer figure,
þat is one, & þat wul be foure. do away þe figure of 3, and write
þere a figure of foure. and lete þe neþer figure stonde stil, & þan
worch forth. vnde versus.

¶ **Articulus si sit, in primo limite cifram,**
¶ **Articulum vero reliquis inscribe figuris,**
Vel per se scribas si nulla figura sequatur.

¶ Here he puttes þe thryde case of þe craft of Addicioun. & þe
case is þis. yf of Addicioun of 2 figuris a-ryse an Articulle, how *Suppose it is an Article,*
schal þou do. thou most do away þe heer figure þat was addid to *set down a ciphe- and*
þe neþer, & write þere a cifre, and sett þe articuls on þe figuris *carry the tens.*
hede, yf þat þere come ony after. And wyrch þan as I haue tolde
þe in þe secunde case. An ensampull 25. Cast 5 to 5, þat wylle
be ten. now do away þe hyer 5, & $\frac{15}{}$ write þere a cifer. And
sette ten vpon þe figuris hed of 2. And reken it but for on þus. Io

The Craft of Subtraction.

¹ leaf 142 a.
Here is an example.

an Ensampulle ⌈1 ⌉ And ¹þan worch forth. But yf þere come no
figure after þe ⌊2ϕ⌋ cifre, write þe articul next hym in þe same rewe
              ⌊15⌋
as here ⌈5⌉.   cast 5 to 5, and it wel be ten. do away 5. þat is þe
hier 5. ⌊5⌋ and write þere a cifre, & write after hym þe articul as 4
þus ⌈1ϕ⌉. And þan þou hast done.
    ⌊5 ⌋

¶ Si tibi cifra superueniens occurrerit, illa*m*
   Dele superposita*m*; fac illic scribe figura*m*,   8
   Postea procedas reliquas addendo figuras.

What to do when you have a cipher in the top row.

¶ Here he puttes þe fourt case, & it is þis, þat yf þere come a
cifer in þe hier rewe, how þou schal do. þus þou schalt do. do
away þe cifer, & sett þere þe digit þat comes of þe addicioun as þus 12

An example of all the difficulties.

1ϕϕ84. In þis ensampul ben alle þe foure cases. Cast 3 to foure,
 17743 þat wol be seuen. do away 4. & write þere seuen; þan cast
4 to þe figure of 8. þat wel be 12. do away 8, & sett þere 2. þat is
a digit, and sette þe articul of þe composit, þat is ten, vpon þe cifers 16
hed, & reken it for hym selfe þat is on. þan cast one to a cifer, &
hit wulle be but on, for noȝt & on makes but one. þan cast 7. þat
stondes vnder þat on to hym, & þat wel be 8. do away þe cifer &
þat 1. & sette þere 8. þan go forthermore. cast þe oþer 7 to þe cifer 20
þat stondes ouer hym. þat wul be bot seuen, for þe cifer betokens

² leaf 142 b.

noȝt. do away þe cifer & sette þere seuen, ²& þen go forþermore
& cast 1 to 1, & þat wel be 2. do away þe hier 1, & sette þere 2.
þan hast þou do. And yf þou haue wel ydo þis nomber þat is sett 24
here-after wel be þe nomber þat schalle aryse of alle þe addicion as
here 27827. ¶ Sequit*ur* alia spec*ies*.

A nu*me*ro numer*um* si sit tibi demere cura
  Scribe figurar*um* series, ʋt in addicione.   28

Four things to know about subtraction:

¶ This is þe Chapter of subtraccion, in the quych þou most
know foure nessessary thynges. the first what is subtraccion. þe
secunde is how mony nombers þou most haue to subtraccion, the
thryd is how many maners of cases þere may happe in þis craft of 32
subtraccion. The fourte is qwat is þe profet of þis craft. ¶ As for

the first;

þe first, þou most know þat subtraccion is drawynge of one

the second;

nowmber oute of anoþer nomber. As for þe secunde, þou most
knowe þat þou most haue two rewes of figuris one vnder anoþer, as 36

the third;

þou addyst in addicion. As for þe thryd, þou moyst know þat
foure maner of diuerse casis mai happe in þis craft. ¶ As for þe

the fourth.

fourt, þou most know þat þe profet of þis craft is whenne þou hasse
taken þe lasse nomber out of þe more to telle what þere leues ouer 40

The Cases of the Craft of Subtraction.

þat. & þou most be-gynne to wyrch in þis craft in þe ryght side of þe boke, as þou diddyst in addicioṅ. Versus.

¶ **Maiori numero numerum suppone minorem,**
4 ¶ **Siue pari numero supponatur numerus par.**

¹ ¶ Here he telles þat þe hier nomber most be more þen þe neþer, ¹ leaf 143 a.
or els eueṅ as mych. but he may not be lasse. And þe case is Put the greater
þis, þou schalt drawe þe neþer nomber out of þe hyer, & þou mayst number above the
8 not do þat yf þe hier nomber were lasse þan þat. ffor þou mayst not less.
draw sex out of 2. But þou mast draw 2 out of sex. And þou
maiste draw twene out of twene, for þou schal leue noȝt of þe hier
twene vnde versus.

12 ¶ **Postea si possis a prima subtrahe primam**
   Scribens quod remanet.

Here is þe first case put of subtraccioṅ, & he says þou schalt The first case of subtrac-
begynne in þe ryght side, & draw þe first figure of þe neþer rewe tion.
16 out of þe first figure of þe hier rewe. qwether þe hier figure be more
þen þe neþer, or eueṅ as mych. And þat is notified in þe vers when
he says " Si possis." Whan þou has þus ydo, do away þe hiest
figure & sett þere þat leues of þe subtraccioṅ, lo an Ensampulle Here is an example.
20 ┌─────┐   draw 2 out of 4. þan leues 2. do away 4 & write þere 2, &
   │ 234 │
   │ 122 │ latte þe neþer figure stonde stille, & so go for-by oþer figuris
   └─────┘
till þou come to þe ende, þan hast þou do.

¶ **Cifram si nil remanebit.**

24 ¶ Here he puttes þe secunde case, & hit is þis. yf it happe þat Put a cipher if nothing
qwen þou hast draw on neþer figure out of a hier, & þere leue noȝt remains.
after þe subtraccioṅ, þus ² þou schalt do. þou schalle do away þe hier ² leaf 143 b.
figure & write þere a cifer, as lo an Ensampull ┌────┐ Take foure Here is an example.
                                                 │ 24 │
28 out of foure þan leus noȝt. þerefore do away │ 24 │ þe hier 4 &
                                                 └────┘
set þere a cifer, þan take 2 out of 2, þan leues noȝt. do away þe
hier 2, & set þere a cifer, and so worch whare so euer þis happe.

Sed si non possis a prima demere primam
32   **Precedens vnum de limite deme sequente,**
      Quod demptum pro denario reputabis ab illo
      Subtrahe totalem numerum quem proposuisti
      Quo facto scribe super quicquid remanebit.

36 Here he puttes þe thryd case, þe quych is þis. yf it happe þat Suppose you cannot take
þe neþer figure be more þen þe hier figure þat he schalle be draw out the lower figure from
of. how schalle þou do. þus þou schalle do. þou schalle borro .1. the top one, borrow ten;
oute of þe next figure þat comes after in þe same rewe, for þis case
40 may neuer happ but yf þere come figures after. þan þou schalt sett

The Cases of the Craft of Subtraction.

take the lower number from ten;
add the answer to the top number.
¹ leaf 144 a.

þat on ouer þe hier figures hed, of the quych þou woldist y-draw oute þe neyþer figure yf þou haddyst y-myʒt. Whane þou hase þus ydo þou schalle rekene þat .1. for ten. ¶. And out of þat ten þou schal draw þe neyþermost figure, And alle þat leues þou schalle adde to þe figure on whos hed þat .1. stode. And þen þou schalle do away alle þat, & sett þere alle that arisys of the addicion of þe ylke 2 figuris. And yf yt ¹happe þat þe figure of þe quych þou schalt borro on be hym self but 1. If þou schalt þat one & sett it vppon þe oþer figuris hed, and sett in þat 1. place a cifer, yf þere

Example. come mony figures after. lo an Ensampul. ⌈2122⌉. take 4 out of 2. it wyl not be, þerfore borro one of þe next ⌊1134⌋ figure, þat is 2. and sett þat ouer þe hed of þe fyrst 2. & rekene it for ten. and þere þe 12 secunde stondes write 1. for þou tokest on out of hym. þan take þe neþer figure, þat is 4, out of ten. And þen leues 6. cast to 6 þe figure of þat 2 þat stode vnder þe hedde of 1. þat was borwed & rekened for ten, and þat wylle be 8. do away þat 6 & þat 2, & sette þere 8, & lette þe neþer figure stonde stille. Whanne þou hast

How to 'Pay back' the borrowed ten.
do þus, go to þe next figure þat is now bot 1. but first yt was 2, & þere-of was borred 1. þan take out of þat þe figure vnder hym, þat is 3. hit wel not be. þer-fore borowe of the next figure, þe quych is 20 bot 1. Also take & sett hym ouer þe hede of þe figure þat þou woldest haue y-draw oute of þe nether figure, þe quych was 3. & þou myʒt not, & rekene þat borwed 1 for ten & sett in þe same place, of þe quych place þou tokest hym of, a cifer, for he was bot 1.

² leaf 144 b.
Whanne þou hast þus ydo, take out of þat 1. þat is rekent for ten, þe neþer figure of 3. And þere leues 7. ²cast þe ylke 7 to þe figure þat had þe ylke ten vpon his hed, þe quych figure was 1, & þat wol be 8. þan do away þat 1 and þat 7, & write þere 8. & þan wyrch forth in oþer figuris til þou come to þe ende, & þan þou hast þe do. Versus.

4
8
12
16
24
28
32

¶ **Facque nonenarios de cifris, cum remeabis**
¶ **Occurrant si forte cifre; dum dempseris vnum**
¶ **Postea procedas reliquas demendo figuras.**

A very hard case is put.
¶ Here he puttes þe fourte case, þe quych is þis, yf it happe þat þe neþer figure, þe quych þou schalt draw out of þe hier figure be more þan þe hier figur ouer hym, & þe next figure of two or of thre or of foure, or how mony þere be by cifers, how wold þou do. þou wost wel þou most nede borow, & þou mayst not borow of þe cifers, for þai haue noʒt þat þai may lene or spare. Ergo³ how

36

³ Perhaps "So."

How to prove the Subtraction.

woldest þou do. Certayn þus most þou do, þou most borow on of
þe next figure significatyf in þat rewe, for þis case may not happe,
but yf þere come figures significatyf after the cifers. Whan þou
4 hast borowede þat 1 of the next figure significatyf, sett þat on ouer
þe hede of þat figure of þe quych þou wold haue draw þe neþer
figure out yf þou hadest my3t, & reken it for ten as þou diddest
in þe oþer case here-a-fore. Whan þou hast þus y-do loke how
8 mony cifers þere were bye-twene þat figure significatyf, & þe figure
of þe quych þou woldest haue y-draw the [1]neþer figure, and of euery   [1] leaf 145 a.
of þe ylke cifers make a figure of 9. lo an Ensampulle after. ⌈40002⌉ Here is an
Take 4 out of 2. it wel not be. borow 1 out of þe next figure ⌊10004⌋ example.
12 significatyf, þe quych is 4, & þen leues 3. do away þat figure of 4
& write þere 3. & sett þat 1 vppon þe figure of 2 hede, & þan take
4 out of ten, & þan þere leues 6. Cast 6 to the figure of 2, þat wol
be 8. do away þat 6 & write þere 8. Whan þou hast þus y-do
16 make of euery 0 betweyn 3 & 8 a figure of 9, & þan worch forth in
goddes name. & yf þou hast wel y-do þou[2] schalt haue þis nomber

¶ **Si subtraccio sit bene facta probare valebis**   ⌈39998⌉ Sic.
   Quas subtraxisti primas addendo figuras.   ⌊10004⌋

20   ¶ Here he teches þe Craft how þou schalt know, whan þou hast How to prove
subtrayd, wheþer þou hast wel ydo or no. And þe Craft is þis, a subtraction sum.
ryght as þou subtrayd þe neþer figures fro þe hier figures, ry3t so
adde þe same neþer figures to þe hier figures. And yf þou haue
24 well y-wroth a-fore þou schalt haue þe hier nombre þe same þou
haddest or þou be-gan to worch. as for þis I bade þou schulde
kepe þe neþer figures stylle. lo an [3]Ensampulle of alle þe 4 cases [3] leaf 145 b.
togedre. worche welle þis case ⌈40003468⌉. And yf þou worch welle Here is an
28 whan þou hast alle subtrayd ⌊20004664⌋ þe þat hier nombre here, example.
þis schalle be þe nombre here foloyng whan þou hast subtrayd
⌈39998804⌉. And þou schalt know þus. adde þe neþer rewe of þe Our author
⌊20004664⌋ same nombre to þe hier rewe as þus, cast 4 to 4. þat wol makes a slip here (3 for 1).
32 be 8. do away þe 4 & write þere 8. by þe first case of addicion.
þan cast 6 to 0 þat wol be 6. do away þe 0, & write þere 6. þan
cast 6 to 8, þat wel be 14. do away 8 & write þere a figure of 4,
þat is þe digit, and write a figure of 1. þat schall be-token ten. þat
36 is þe articul vpon þe hed of 8 next after, þan reken þat 1. for 1. &
cast it to 8. þat schal be 9. cast to þat 9 þe neþer figure vnder þat
þe quych is 4, & þat schalle be 13. do away þat 9 & sett þere 3, &
sett a figure of 1. þat schall be 10 vpon þe next figuris hede þe

       [2] 'hali' marked for erasure in MS.

quych is 9. by þe secunde case þat þou hadest in addicion). þan cast
1 to 9. & þat wol be 10. do away þe 9. & þat 1. And write þere a
cifer. and write þe articulle þat is 1. betokenynge 10. vpon þe hede of
þe next figure toward þe lyft side, þe quych is 9, & so do forth tyl
þou come to þe last 9. take þe figure of þat 1. þe quych þou schalt
fynde ouer þe hed of 9. & sett it ouer þe next figures hede þat
schal be 3. ¶ Also do away þe 9. & set þere a cifer, & þen cast
þat 1 þat stondes vpon þe hede of 3 to þe same 3, & þat schalle make
4, þen caste to þe ylke 4 the figure in þe neyþer rewe, þe quych is
2, and þat schalle be 6. And þen schal þou haue an Ensampulle
aȝeyn), loke & se, & but þou haue þis same þou hase myse-wroȝt.

Sequitur de duplacione

Si vis duplare numerum, sic incipe primo
Scribe figurarum seriem quamcunque velis tu.

¶ This is the Chapture of duplacion), in þe quych craft þou most
haue & know 4 thinges. ¶ þe first þat þou most know is what is
duplacion). þe secunde is how mony rewes of figures þou most
haue to þis craft. ¶ þe thryde is how many cases may happe in
þis craft. ¶ þe fourte is what is þe profet of þe craft. ¶ As for þe
first. duplacion) is a doublynge of a nombre. ¶ As for þe secunde
þou most haue on' nombre or on rewe of figures, the quych called
numerus duplandus. As for þe thrid þou most know þat 3 diuerse
cases may hap in þis craft. As for þe fourte. qwat is þe profet of
þis craft, & þat is to know what a-risyȝt of a nombre I-doublyde.
¶ fforþer-more, þou most know & take gode hede in quych side þou
schalle be-gyn in þis craft, or ellis þou mayst spyl alle þi laber þere
aboute. certeyn þou schalt begyn) in the lyft side in þis Craft.
thenke wel ouer þis verse. ¶ A leua dupla, diuide, multiplica.
The sentens of þes verses afore, as þou may see if þou take hede.
As þe text of þis verse, þat is to say, ¶ Si vis duplare. þis is þe
sentence. ¶ If þou wel double a nombre þus þou most be-gynn).
Write a rewe of figures of what nombre þou welt. versus.

Postea procedas primam duplando figuram
Inde quod excrescit scribas vbi iusserit ordo
Iuxta precepta tibi que dantur in addicione.

¶ Here he telles how þou schalt worch in þis Craft. he says,
fyrst, whan þou hast writen þe nombre þou schalt be-gyn at þe first

---

[2] 'moy' in MS.
[4] Subtrahas aut addis a dextris vel mediabis' added on margin of MS.

figure in the lyft side, & doubulle þat figure, & þe nombre þat comes
þere-of þou schalt write as þou diddyst in addicioŋ, as ¶ I schal telle
þe in þe case. versus.

4   ¹ ¶ **Nam si sit digitus in primo limite scribas.**   ¹ leaf 147 a.

¶ Here is þe first case of þis craft, þe quych is þis. yf of dupla- If the answer is a digit,
cioŋ of a figure arise a digit. what schal þou do. þus þou schal
do. do away þe figure þat was doublede, & sett þere þe diget þat write it in the place of
8 comes of þe duplacioŋ, as þus. 23. double 2, & þat wel be 4. do the top figure.
away þe figure of 2 & sett þere a figure of 4, & so worch forth tille
þou come to þe ende. versus.

    ¶ **Articulus si sit, in primo limite cifram,**
12   ¶ **Articulum vero reliquis inscribe figuris;**
    ¶ **Vel per se scribas, si nulla figura sequatur.**

¶ Here is þe secunde case, þe quych is þis yf þere come an If it is an article,
articulle of þe duplacioŋ of a figure þou schalt do ryȝt as þou
16 diddyst in addicioŋ, þat is to wete þat þou schalt do away þe
figure þat is doublet & sett þere a cifer, & write þe articulle ouer þe put a cipher in the place,
next figuris hede, yf þere be any after-warde toward þe lyft side as and 'carry' the tens.
þus. 25. begyn at the lyft side, and doubulle 2. þat wel be 4. do
20 away þat 2 & sett þere 4. þan doubul 5. þat wel be 10. do away 5,
& sett þere a 0, & sett 1 vpon þe next figuris hede þe quych is 4.
& þen draw downe 1 to 4 & þat wolle be 5, & þen do away þat 4
& þat 1, & sett þere 5. for þat 1 schal be rekened in þe drawynge to-
24 gedre for 1. wen ² þou hast ydon þou schalt haue þis nombre 50.   ² leaf 147 b.
yf þere come no figure after þe figure þat is addit, of þe quych If there is no figure to
addicioŋ comes an articulle, þou schalt do away þe figure þat is 'carry' them to, write
dowblet & sett þere a 0. & write þe articul next by in þe same them down.
28 rewe toward þe lyft syde as þus, 523. double 5 þat woll be ten. do
away þe figure 5 & set þere a cifer, & sett þe articul next after in
þe same rewe toward þe lyft side, & þou schalt haue þis nombre
1023. þen go forth & double þe oþer nombers þe quych is lyȝt y-
32 nowȝt to do. versus.

    ¶ **Compositus si sit, in limite scribe sequente**
    Articulum, primo digitum; quia sic iubet ordo:
    Et sic de reliquis faciens, si sint tibi plures.

36   ¶ Here he puttes þe Thryd case, þe quych is þis, yf of dupla- If it is a Composite,
cioŋ of a figure come a Composit. þou schalt do away þe figure þat
is doublet & set þere a digit of þe Composit, & sett þe articulle ouer write down the digit,
þe next figures hede, & after draw hym downe with þe figure ouer and 'carry' the tens
40 whos hede he stondes, & make þere-of an nombre as þou hast done

The Craft of Mediation.

¹ leaf 148 a.
Here is an example.

afore, & yf þere come no figure after þat digit þat þou hast y-write, þan set þe articulle next after hym in þe same rewe as þus, 67 : double 6 þat wel be 12, do away 6 & write þere þe digit [1] of 12, þe quych is 2, and set þe articulle next after toward þe lyft side in þe same rewe, for þere comes no figure after. þan dowble þat oþer figure, þe quych is 7, þat wel be 14. the quych is a Composit. þen do away 7 þat þou doublet & sett þe þe diget of hym, the quych is 4, sett þe articulle ouer þe next figures hed, þe quych is 2, & þen draw to hym þat on, & make on nombre þe quych schalle be 3. And þen yf þou haue wel y-do þou schalle haue þis nombre of þe duplacioun, 134. versus. 4

8

¶ Si super extremam nota sit monadem dat eidem
Quod tibi contingat si primo dimidiabis. 12

How to double the mark for one-half.

¶ Here he says, yf ouer þe fyrst figure in þe ryȝt side be such a merke as is here made, ˮ, þou schalle fyrst doubulle þe figure, the quych stondes vnder þat merke, & þen þou schalt doubul þat merke þe quych stondes for haluendel on. for too haluedels makes on, & so þat wol be on. cast þat on to þat duplacioun of þe figure ouer whos hed stode þat merke, & write it in þe same place þere þat þe figure þe quych was doublet stode, as þus 23ˮ. double 3, þat wol be 6 ; doubul þat halue on, & þat wol be on. cast on to 6, þat wel be 7. do away 6 & þat 1, & sett þere 7. þan hase þou do. as for þat figure, þan go [2] to þe oþer figure & worch forth. & þou schall neuer haue such a merk but ouer þe hed of þe furst figure in þe ryght side. And ȝet it schal not happe but yf it were y-halued a-fore, þus þou schalt vnderstonde þe verse. ¶ Si super extremam &c. Et nota, talis figura ˮ significans medietatem, unitatis veniat, i.e. contingat uel fiat super extremam, i.e. super primam figuram in extremo sic versus dextram ars dat : i.e. reddit monadem. i.e. vnitatem eidem. i.e. eidem note & declina tur hec monos, dis, di, dem, &c. ¶ Quod ergo totum hoc dabis monadem note continget. i.e. eveniet tibi si dimidiasti, i.e. accipisti uel subtulisti medietatem alicuius unius, in cuius principio sint figura numerum denotans imparem primo i.e. principiis. 16

20

² leaf 148 b.
This can only stand over the first figure.

24

28

32

¶ Sequitur de mediacione.

Incipe sic, si vis aliquem numerum mediare :
Scribe figurarum seriem solam, velut ante.

The four things to be known in mediation:

¶ In þis Chapter is taȝt þe Craft of mediacioun, in þe quych craft þou most know 4 thynges. ffurst what is mediacion. the secunde how many rewes of figures þou most haue in þe wyrchynge of þis craft. þe thryde how many diuerse cases may happ in þis craft.[3] ¶ As for þe furst, þou schalt vndurstonde þat mediacion is a 36

the first

40

³ After 'craft' insert 'the .4. what is þe profet of þis craft.'

The Mediation of an Odd Number.   17

takyng out of halfe a nomber out of a holle nomber, [1]as yf þcu   [1] leaf 149 a.
wolde take 3 out of 6. ¶ As for þe secunde, þou schalt know þat the second;
þou most haue one rewe of figures, & no moo, as þou hayst in þe
4 craft of duplacion). ¶ As for the thryd, þou most vnderstonde þat the third;
5 cases may happe in þis craft. ¶ As for þe fourte, þou schalle the fourth.
know þat the profet of þis craft is when þou hast take away þe
haluendel of a nombre to telle qwat þere schalle leue. ¶ Incipe
8 sic, &c. The sentence of þis verse is þis. yf þou wold medye, þat
is to say, take halfe out of þe holle, or halfe out of halfe, þou most
begynne þus. Write one rewe of figures of what nombre þou wolte, Begin thus.
as þou dyddyst be-fore in þe Craft of duplacion). versus.
12   ¶ **Postea procedas medians, si prima figura**
    Si par aut impar videas.

¶ Here he says, when þou hast write a rewe of figures, þou
schalt take hede wheþer þe first figure be euen) or odde in nombre, See if the
16 & vnderstonde þat he spekes of þe first figure in þe ry3t side. And number is even or odd.
in the ryght side þou schalle begynne in þis Craft.
    ¶ **Quia si fuerit par,**
  Dimidiabis eam, scribens quicquid remanebit:
20   ¶ Here is the first case of þis craft, þe quych is þis, yf þe first If it is even,
figure be euen. þou schal take away fro þe figure euen halfe, & do halve it, and write the
away þat figure and set þere þat leues ouer, as þus, 4. take [2]halfe answer in its place.
out of 4, & þan þere leues 2. do away 4 & sett þere 2. þis is lyght   [2] leaf 149 b.
24 y-now3t. versus.
  ¶ **Impar si fuerit vnum demas mediare**
    Quod non presumas, sed quod superest mediabis
    Inde super tractum fac demptum quod notat vnum.
28   Here is þe secunde case of þis craft, the quych is þis. yf þe If it is odd,
first figure betokene a nombre þat is odde, the quych odde schal not halve the even number
be mediete, þen þou schalt medye þat nombre þat leues, when the less than it.
odde of þe same nombre is take away, & write þat þat leues as þou
32 diddest in þe first case of þis craft. Whan) þou hayst write þat. for
þat þat leues, write such a merke as is here ͫ vpon his hede, þe quych Then write
merke schal betoken) halfe of þe odde þat was take away. lo an the sign for one-half over it.
Ensampull. 245. the first figure here is betokenynge odde nombre,
36 þe quych is 5, for 5 is odde; þere-fore do away þat þat is odde, þe Here s an example.
quych is 1. þen leues 4. þen medye 4 & þen leues 2. do away 4. &
sette þere 2, & make such a merke ͫ upon his hede, þat is to say
ouer his hede of 2 as þus. 242.ͫ And þen worch forth in þe oþer
40 figures tyll þou come to þe ende. by þe furst case as þou schalt
  NOMBRYNGE.   C
3 ★

The Cases of the Craft of Mediation.

margin: ¹ leaf 150 a. Put the mark only over the first figure.

vnderstonde þat þou schalt ¹neuer make such a merk but ouer þe first figure hed in þe riȝt side. Wheþer þe other figures þat comyn after hym be euen or odde. versus.

    ¶ **Si monos, dele; sit** tibi **cifra post** nota **supra.**   4

margin: If the first figure is one put a cipher.

¶ Here is þe thryde case, þe quych yf the first figure be a figure of 1. þou schalt do away þat 1 & set þere a cifer, & a merke ouer þe cifer as þus, 241. do away 1, & sett þere a cifer with a merke ouer his hede, & þen hast þou ydo for þat 0. as þus 0⁔ þen worch forth 8 in þe oþer figurys till þou come to þe ende. for it is lyght as dyche water. vnde versus.

    ¶ **Postea** procedas **hac condic**ione **secu**nda:
    Impar si fuerit **hinc vnu**m **deme pr**iori,   12
    Inscribens quinque, nam denos significabit
    Monos predict**am.**

margin: What to do if any other figure is odd.

¶ Here he puttes þe fourte case, þe quych is þis. yf it happen the secunde figure betoken odde nombre, þou schal do away on of 16 þat odde nombre, þe quych is significatiue by þat figure. 1. þe quych 1 schall be rekende for 10. Whan þou hast take away þat 1 out of þe nombre þat is significde by þat figure, þou schalt medie þat þat leues ouer, & do away þat figure þat is medied, & sette in his styde 20 halfe of þat nombre. ¶ Whan þou hase so done, þou schalt write

margin: ² leaf 150 b. Write a figure of five over the next lower number's head.

²a figure of 5 ouer þe next figures hede by-fore toward þe ryȝt side, for þat 1, þe quych made odd nombre, schall stonde for ten, & 5 is halfe of 10; so þou most write 5 for his haluendelle. lo an En- 24 sampulle, 4678. begyn in þe ryȝt side as þou most nedes. medie 8.

margin: Example.

þen þou schalt leue 4. do away þat 8 & sette þere 4. þen out of 7. take away 1. þe quych makes odde, & sett 5. vpon þe next figures hede afore toward þe ryȝt side, þe quych is now 4. but afore it was 28 8. for þat 1 schal be rekenet for 10, of þe quych 10, 5 is halfe, as þou knowest wel. Whan þou hast þus ydo, medye þat þe quych leues after þe takyinge away of þat þat is odde, þe quych leuynge schalle be 3; do away 6 & sette þere 3, & þou schalt haue such a 32 nombre 463⁵4. after go forth to þe next figure, & medy þat, & worch forth, for it is lyȝt ynovȝt to þe certayn.

    ¶ **Si v**ero **secu**nda **dat vnu**m.
    Illa deleta, scri**batur cifra; pr**iori   36
    ¶ **Tradendo quinque pro denario mediato;**
    Nec cifra scri**batur, nisi dei**nde **figu**ra **sequat**ur:
    Postea procedas **reliqu**as **mediando figuras**
    Vt supra docui, si sint tibi mille figure.   40

How to prove the Mediation.

¶ Here he puttes þe 5 case, þe quych is [1]þis : yf þe secunde figure be of 1, as þis is here 12, þou schalt do away þat 1 & sett þere a cifer. & sett 5 ouer þe next figure hede afore toward þe riȝt 4 side, as þou diddyst afore; & þat 5 schal be haldel of þat 1, þe quych 1 is rekent for 10. lo an Ensampulle, 214. medye 4. þat schalle be 2. do away 4 & sett þere 2. þen go forth to þe next figure. þe quych is bot 1. do away þat 1. & sett þere a cifer. & set 8 5 vpon þe figures hed afore, þe quych is nowe 2, & þen þou schalt haue þis nombre 202,⁵ þen worch forth to þe nex figure. And also it is no maystery yf þere come no figure after þat on is medyet, þou schalt write no 0. ne nowȝt ellis, but set 5 ouer þe next figure afore 12 toward þe ryȝt, as þus 14. medie 4 then leues 2, do away 4 & sett þere 2. þen medie 1. þe quich is rekende for ten, þe haluendel þere-of wel be 5. sett þat 5 vpon þe hede of þat figure, þe quych is now 2, & do away þat 1, & þou schalt haue þis nombre yf þou 16 worch wel, 2⁵. vnde versus.

Marginal notes: [1] leaf 151 a. If the second figure is one, put a cipher, and write five over the next figure. How to halve fourteen.

¶ **Si mediacio sit bene facta probare valebis**
¶ **Duplando numerum quem primo dimediasti**

¶ Here he telles þe how þou schalt know wheþer þou hase wel 20 ydo or no. doubul [2]þe nombre þe quych þou hase mediet, and yf þou haue wel y-medyt after þe dupleacion, þou schalt haue þe same nombre þat þou haddyst in þe tabulle or þou began to medye, as þus. ¶ The furst ensampulle was þis. 4. þe quych I-mediet was 24 laft 2, þe whych 2 was write in þe place þat 4 was write afore. Now doubulle þat 2, & þou schal haue 4, as þou hadyst afore. þe secunde Ensampulle was þis, 245. When þou haddyst mediet alle þis nombre, yf þou haue wel ydo þou schalt haue of þat mediacion 28 þis nombre, 122ʷ. Now doubulle þis nombre, & begyn in þe lyft side; doubulle 1, þat schal be 2. do away þat 1 & sett þere 2. þen doubulle þat oþer 2 & sett þere 4, þen doubulle þat oþer 2, & þat wel be 4. þen doubul þat merke þat stondes for halue on. & þat schalle 32 be 1. Cast þat on to 4, & it schalle be 5. do away þat 2 & þat merke, & sette þere 5, & þen þou schal haue þis nombre 245. & þis wos þe same nombur þat þou haddyst or þou began to medye, as þou mayst se yf þou take hede. The nombre þe quych þou haddist 36 for an Ensampul in þe 3 case of mediacion to be mediet was þis 241. whan þou haddist medied alle þis nombur truly [3]by euery figure, þou schall haue be þat mediacion þis nombur 120ʷ. Now dowbul þis nombur, & begyn in þe lyft side, as I tolde þe in þe 40 Craft of duplacion. þus doubulle þe figure of 1, þat wel be 2. do

Marginal notes: How to prove your mediation. [2] leaf 151 b. First example. The second. The third example. [3] leaf 152 a.

The Craft of Multiplication.

away þat 1 & sett þere 2, þen doubul þe next figure afore, the quych is 2, & þat wel be 4; do away 2 & set þere 4. þen doubul þe cifer, & þat wel be noȝt, for a 0 is noȝt. And twyes noȝt is but noȝt. þerefore doubul the merke aboue þe cifers hede, þe quych be- 4 tokenes þe haluendel of 1, & þat schal be 1. do away þe cifer & þe merke, & sett þere 1, & þen þou schalt haue þis nombur 241. And þis same nombur þou haddyst afore or þou began to medy, & yf þou take gode hede. ¶ The next ensampul þat had in þe 4 case 8 of mediacion was þis 4678. Whan þou hast truly ymedit alle þis nombur fro þe begynnynge to þe endynge, þou schalt haue of þe mediacion þis nombur 2334. Now doubul this nombur & begyn in þe lyft side, & doubulle 2 þat schal be 4. do away 2 and sette þere 12 4; þen doubule 3, þat wol be 6; do away 3 & sett þere 6, þen doubul þat oþer 3, & þat wel be 6; do away 3 & set þere [1]6, þen doubul þe 4, þat welle be 8; þen doubul 5. þe quych stondes ouer þe hed of 4, & þat wol be 10; cast 10 to 8, & þat schal be 18; do 16 away 4 & þat 5, & sett þere 8, & sett that 1, þe quych is an articul of þe Composit þe quych is 18, ouer þe next figures hed toward þe lyft side, þe quych is 6. drav þat 1 to 6, þe quych 1 in þe dravyng schal be rekente bot for 1, & þat 1 & þat 6 togedur wel be 7. do 20 away þat 6 & þat 1. the quych stondes ouer his hede, & sett ther 7, & þen þou schalt haue þis nombur 4678. And þis same nombur þou hadyst or þou began to medye, as þou mayst see in þe secunde Ensampul þat þou had in þe 4 case of mediacion, þat was þis: when 24 þou had mediet truly alle the nombur, a principio usque ad finem. þou schalt haue of þat mediacion þis nombur 102. Now doubul 1. þat wel be 2. do away 1 & sett þere 2. þen doubul 0. þat will be noȝt. þerefore take þe 5, þe quych stondes ouer þe next figures 28 hed, & doubul it, & þat wol be 10. do away þe 0 þat stondes betwene þe two figuris, & sette þere in his stid 1, for þat 1 now schal stonde in þe secunde place, where he schal betoken 10; þen doubul 2, þat wol be 4. do away 2 & sett þere 4. & [2]þou schal haue 32 þus nombur 214. þis is þe same numbur þat þou hadyst or þou began to medye, as þou may see. And so do euer more, yf þou wil knowe wheþer þou hase wel ymedyt or no. ¶. doubulle þe numbur þat comes after þe mediacioun, & þou schal haue þe same nombur 36 þat þou hadyst or þou began to medye, yf þou haue welle ydo. or els doute þe noȝt, but yf þou haue þe same, þou hase faylide in þi Craft.

Sequitur de multiplicatione. 40

To write down a Multiplication Sum.

Si tu per numerum numerum vis multiplicare
Scribe duas quascunque velis series numerorum
Ordo servetur vt vltima multiplicandi
Ponatur super anteriorem multiplicantis
A leua relique sint scripte multiplicantes.

¶ Here be-gynnes þe Chaptre of multiplicatiou), in þe quych þou most know 4 thynges. ¶ Ffirst, qwat is multiplicacion). The secunde, how mony cases may hap in multiplicacion). The thryde, how mony rewes of figures þere most be. ¶ The 4. what is þe profet of þis craft. ¶ As for þe first, þou schal vnderstonde þat multiplicacion) is a bryngynge to-geder of 2 thynges in on nombur, þe quych on nombur contynes so mony tymes on, howe ¹mony tymes þere ben vnytees in þe nowmbre of þat 2, as twyes 4 is 8. now here ben þe 2 nombers, of þe quych too nowmbres on is betokened be an aduerbe, þe quych is þe worde twyes, & þis worde thryes, & þis worde foure sythes,² & so furth of such other lyke wordes. ¶ And tweyn nombres schal be tokenyde be a nowne, as þis worde foure showys þes tweyn) nombres y-broth in-to on hole nombur, þat is 8, for twyes 4 is 8, as þou wost wel. ¶ And þes nombre 8 conteynes as oft tymes 4 as þere ben vnites in þat other nombre, þe quych is 2, for in 2 ben 2 vnites, & so oft tymes 4 ben in 8, as þou wottys wel. ¶ Ffor þe secunde, þou most know þat þou most haue too rewes of figures. ¶ As for þe thryde, þou most know þat 8 maner of diuerse case may happe in þis craft. The profet of þis Craft is to telle when a nombre is multiplyed be a noþer, qwat commys þere of. ¶ Ffforthermore, as to þe sentence of oure verse, yf þou wel multiply a nombur be a-noþer nombur, þou schalt write ³a rewe of figures of what nomburs so euer þou welt, & þat schal be called Numerus multiplicandus, Anglice, þe nombur the quych to be multiplied. þen þou schalt write a-nother rewe of figures, by þe quych þou schalt multiplie the nombre þat is to be multiplied, of þe quych nombur þe furst figure schal be write vnder þe last figure of þe nombur, þe quych is to be multiplied. And so write forthe toward þe lyft side, as here you may se, [67324 / 1234] And þis one nombur schalle be called numerus multi- plicans. Ang- lice, þe nombur multipliynge, for he schalle multiply þe hyer noun-bur, as þus one tyme 6. And so forth, as I schal telle the afterwarde. And þou schal begyn in þe lyft side. ¶ Ffor-þere-more þou schalt vnderstonde þat þere is two manurs of multiplicacion) ; one ys of þe wyrchynge of þe boke only in þe mynde of a mon. fyrst he

Four things to be known of Multiplication:
the first:
¹ leaf 153 b.
the second:
the third:
the fourth.
* leaf 154 a.
The multiplicand.
How to set down the sum.
Two sorts of Multiplication:
mentally,

² After 'sythes' insert '& þis werdes fyue sithe & sex sythes.'

The Craft of Multiplication.

and on paper. teches of þe fyrst maner of duplacion, þe quych is be wyrchynge of tabuls. Afterwarde he wol teche on þe secunde maner. vnde versus.

> **In digitum cures digitum si ducere maior**
> **Per quantum distat a denis respice debes**
> ¶ **Namque suo decuplo totiens delere minorem**
> **Sitque tibi numerus veniens exinde patebit.**

1 leaf 154 b.

How to multiply two digits. ¶ Here he teches a rewle, how þou schalt fynde þe nounbre þat comes by þe multiplicacion of a digit be anoþer. loke how mony [vny]tes ben. bytwene þe more digit and 10. And reken ten for on vnite. And so oft do away þe lasse nounbre out of his owne *Subtract the greater from ten;* decuple, þat is to say, fro þat nounbre þat is ten tymes so mych is þe nounbre þat comes of þe multiplicacion. As yf þou wol multiply 2 be 4. loke how mony vnitees ben by-twene þe quych is þe more nounbre, & be-twene ten. Certen þere wel be vj vnitees by-twene 4 *take the less so many times from ten times itself.* & ten. yf þou reken þere with þe ten þe vnite, as þou may se. so mony tymes take 2. out of his decuple, þe quych is 20. for 20 is þe decuple of 2, 10 is þe decuple of 1, 30 is þe decuple of 3, 40 is þe decuple of 4, And þe oþer digetes til þou come to ten; & whan þou *Example.* hast y-take so mony tymes 2 out of twenty, þe quych is sex tymes, þou schal leue 8 as þou wost wel, for 6 times 2 is twelue. take [1]2 out of twenty, & þere schal leue 8. bot yf bothe þe digettes *2 leaf 155 a.* ²ben y-lyech mych as here. 222 or too tymes twenty, þen it is no fors quych of hem tweyn þou take out of here decuple. als mony tymes as þat is fro 10. but neuer-þe-lesse, yf þou haue hast to *Better use this table, though.* worch, þou schalt haue here a tabul of figures, where-by þou schalt se a-nonn ryght what is þe nounbre þat comes of þe multiplicacion of 2 digittes. þus þou schalt worch in þis figure.

4

8

12

16

20

24

28

```
 1|
 2|  4|
 3|  6|  9|
 4|  8| 12| 16|
 5| 10| 15| 20| 25|
 6| 12| 18| 24| 30| 36|
 7| 14| 21| 28| 35| 42| 49|
 8| 16| 24| 32| 40| 48| 56| 64|
 9| 18| 27| 36| 45| 54| 63| 72| 81|
 1|  2|  3|  4|  5|  6|  7|  8|  9|
```

How to use it. yf þe figure, þe quych schalle be multiplied, be euene as mych as þe diget be, þe quych þat oþer figure schal be multiplied, as two tymes twayn, or thre tymes 3. or sych other. loke qwere þat figure sittes in

29

To multiply one Digit by another. 23

þe lyft side of þe triangle, & loke qwere þe diget sittes in þe neþer
most rewe of þe triangle. & go fro hym vpwarde in þe same rewe,
þe quych rewe gose vpwarde til þou come agaynes þe oþer digette þat
4 sittes in þe lyft side of þe triangle. And þat nounbre, þe quych þou
fyn¹des þere is þe nounbre þat comes of the multiplicacioun of þe 2
digittes, as yf þou wold wete qwat is 2 tymes 2. loke quere sittes
2 in þe lyft side in þe first rewe, he sittes next 1 in þe lyft side al
8 on hye, as þou may se; þe[n] loke qwere sittes 2 in þe lowyst rewe
of þe triangle, & go fro hym vpwarde in þe same rewe tylle þou
come a-ȝenenes 2 in þe hyer place, & þer þou schalt fynd ywrite 4,
& þat is þe nounbre þat comes of þe multiplicacioun of two tymes
12 tweyn is 4, as þow wotest welle. yf þe diget. the quych is multiplied, be more þan þe oþer, þou schalt loke qwere þe more diget
sittes in þe lowest rewe of þe triangle, & go vpwarde in þe same
rewe tyl² þou come a-nendes þe lasse diget in the lyft side. And
16 þere þou schalt fynde þe nombre þat comes of þe multiplicacioun;
but þou schalt vnderstonde þat þis rewle, þe quych is in þis verse.
¶ In digitum cures, &c., noþer þis triangle schalle not serue, bot to
fynde þe nounbres þat comes of the multiplicacioun þat comes of 2
20 articuls or composites, þe nedes no craft but yf þou wolt multiply
in þi mynde. And ³þere-to þou schalt haue a craft afterwarde, for
þou schall wyrch with digettes in þe tables, as þou schalt know
afterwarde. versus.

24   ¶ **Postea procedas postremam multiplicando**
    [Recte multiplicans per cunctas inferiores]
    Condicionem tamen tali quod multiplicantes
    Scribas in capite quicquid processerit inde
28   **Sed postquam fuit hec multiplicate figure**
    Anteriorentur serei multiplicantis
    Et sic multiplica velut isti multiplicasti
    Qui sequitur numerum scriptum quiscunque figuris.

32 ¶ Here he teches how þou schalt wyrch in þis craft. þou schalt
multiplye þe last figure of þe nombre, and quen þou hast so ydo þou
schalt draw alle þe figures of þe neþer nounbre more taward þe ryȝt
side, so qwen þou hast multiplyed þe last figure of þe heyer nounbre
36 by alle þe neþer figures. And sette þe nounbir þat comes þer-of ouer
þe last figure of þe neþer nounbre, & þen þou schalt sette al þe oþer
figures of þe neþer nounbre more nere to þe ryȝt side. ¶ And whan
þou hast multiplied þat figure þat schal be multiplied þe next after

Marginalia: The way to use the Multiplication table. ¹ leaf 155 b. ³ leaf 156 a. How to multiply one number by another. Multiply the 'last' figure of the higher by the 'first' of the lower number.

² 't·l' marked for erasure before 'tyl' in MS.

The Craft of Multiplication.

margin: 1 leaf 156 b.
margin: Set the answer over the first of the lower:
margin: then multiply the second of the lower, and so on.
margin: Then antery the lower number:
margin: as thus.
margin: 2 leaf 157 a.
margin: Now multiply by the last but one of the higher:
margin: as thus.
margin: 4 leaf 157 b.

hym by al þe neþer figures. And worch as þou dyddyst afore til ¹þou come to þe ende. And þou schalt vnderstonde þat euery figure of þe hier nounbre schal be multiplied be alle þe figures of the neþer nounbre, yf þe hier nounbre be any figure þen one. lo an 4 Ensampul here folowynge. ⎡2465⎤. þou schalt begyne to multiplye in þe lyft side. Multiply ⎣232⎦ 2 be 2, and twyes 2 is 4. set 4 ouer þe hed of þat 2, þen multiplie þe same hier 2 by 3 of þe nether nounbre, as thryes 2 þat schal be 6. set 6 ouer þe hed of 3, þan 8 multiplie þe same hier 2 by þat 2 þe quych stondes vnder hym, þat wol be 4 ; do away þe hier 2 & sette þere 4. ¶ Now þou most antery þe nether nounbre, þat is to say, þou most sett þe neþer nounbre more towarde þe ryȝt side, as þus. Take þe neþer 2 toward 12 þe ryȝt side, & sette it euen vnder þe 4 of þe hyer nounbre, & antery alle þe figures þat comes after þat 2, as þus ; sette 2 vnder þe 4. þen sett þe figure of 3 þere þat þe figure of 2 stode, þe quych is now vndur þat 4 in þe hier nounbre ; þen sett þe oþer figure of 2, 16 þe quych is þe last figure toward þe lyft side of þe neþer nomber þere þe figure of 3 stode. þen þou schalt haue such a nombre ⎡464465⎤ ²¶ Now multiply 4, þe quych comes next after 6, by þe last ⎣232⎦ 2 of þe neþer nounbur toward þe lyft side. as 2 tymes 4, þat wel be 20 8. sette þat 8 ouer þe figure the quych stondes ouer þe hede of þat 2, þe quych is þe last figure of þe neþer nounbre ; þan multiplie þat same 4 by 3, þat comes in þe neþer rewe, þat wol be 12. sette þe digit of þe composyt ouer þe figure þe quych stondes ouer þe hed of 24 þat 3, & sette þe articule of þis composit ouer al þe figures þat stondes ouer þe neþer 2 hede. þen multiplie þe same 4 by þe 2 in þe ryȝt side in þe neþer nounbur, þat wol be 8. do away 4. & sette þere 8. Euer more qwen þou multiplies þe hier figure by þat figure 28 þe quych stondes vnder hym, þou schalt do away þat hier figure, & sett þer þat nounbre þe quych comes of multiplicacion of ylke digittes. Whan þou hast done as I haue byde þe, þou schalt haue suych an order of figure as is here, ⎡⎧8 2⎫⎤. þen take and antery 32 þi neþer figures. And sett þe fyrst ⎣⎩4648[65] 232⎭⎦ figure of þe neþer figures³ vndre þe figure of 6. ¶ And draw al þe oþer figures of þe same rewe to hym-warde, ⁴as þou diddyst afore. þen multiplye 6 be 2, & sett þat þe quych comes ouer þere-of 36 ouer al þe oþer figures hedes þat stondes ouer þat 2. þen multiply 6 be 3, & sett alle þat comes þere-of vpon alle þe figures hedes þat standes ouer þat 3 ; þan multiplye 6 be 2, þe quych

---

³ Here 'of þe same rew' is marked for erasure in MS.

To multiply one Composite by another.

stondes vnder þat 6, þen do away 6 & write þere þe digitt of
þe composit þat schal come þereof, & sette þe articull ouer alle
þe figures þat stondes ouer þe hede of þat 3 as here, þen
antery þi figures as þou diddyst afore, and multipli 5
be 2, þat wol be 10; sett þe 0 ouer all þe figures þat
stonden ouer þat 2, & sett þat 1. ouer the next figures
hedes, alle on hye towarde þe lyft side.   þen multiplye 5 be 3. þat
wol be 15, write 5 ouer þe figures hedes þat stonden ouer þat 3, &
sett þat 1 ouer þe next figures hedes toward þe lyft side.   þen
multiplye 5 be 2, þat wol be 10. do away þat 5 & sett þere a 0,
& sett þat 1 ouer þe figures hedes þat stonden ouer 3. And þen
þou schalt haue such a nounbre as here stondes aftur. [1]
¶ Now draw alle þese figures downe togeder as þus, 6.8.1.
& 1 draw to-gedur; þat wolle be 16, do away alle þese
figures saue 6. lat hym stonde, for þow þou take hym
away þou most write þer þe same aȝene. þerefore late
hym stonde, & sett 1 ouer þe figure hede of 4 toward þe lyft side;
þen draw on to 4, þat wolle be 5. do away þat 4 & þat 1, & sette
þere 5.   þen draw 4221 & 1 togedur, þat wol be 10.   do away alle
þat, & write þere þat 4 & þat 0, & sett þat 1 ouer þe next figures
hede toward þe lyft side, þe quych is 6.   þen draw þat 6 & þat 1
togedur, & þat wolle be 7 ; do away 6 & sett þere 7, þen draw 8810
& 1, & þat wel be 18 ; do away alle þe figures þat stondes ouer þe
hede of þat 8, & lette 8 stonde stil, & write þat 1 ouer þe next
figuris hede, þe quych is a 0.   þen do away þat 0, & sett þere 1, þe
quych stondes ouer þe 0. hede.   þen draw 2, 5, & 1 togedur, þat
wolle be 8. þen do away alle þat, & write þere 8. ¶ And þen þou
schalt haue þis nounbre, 571880.

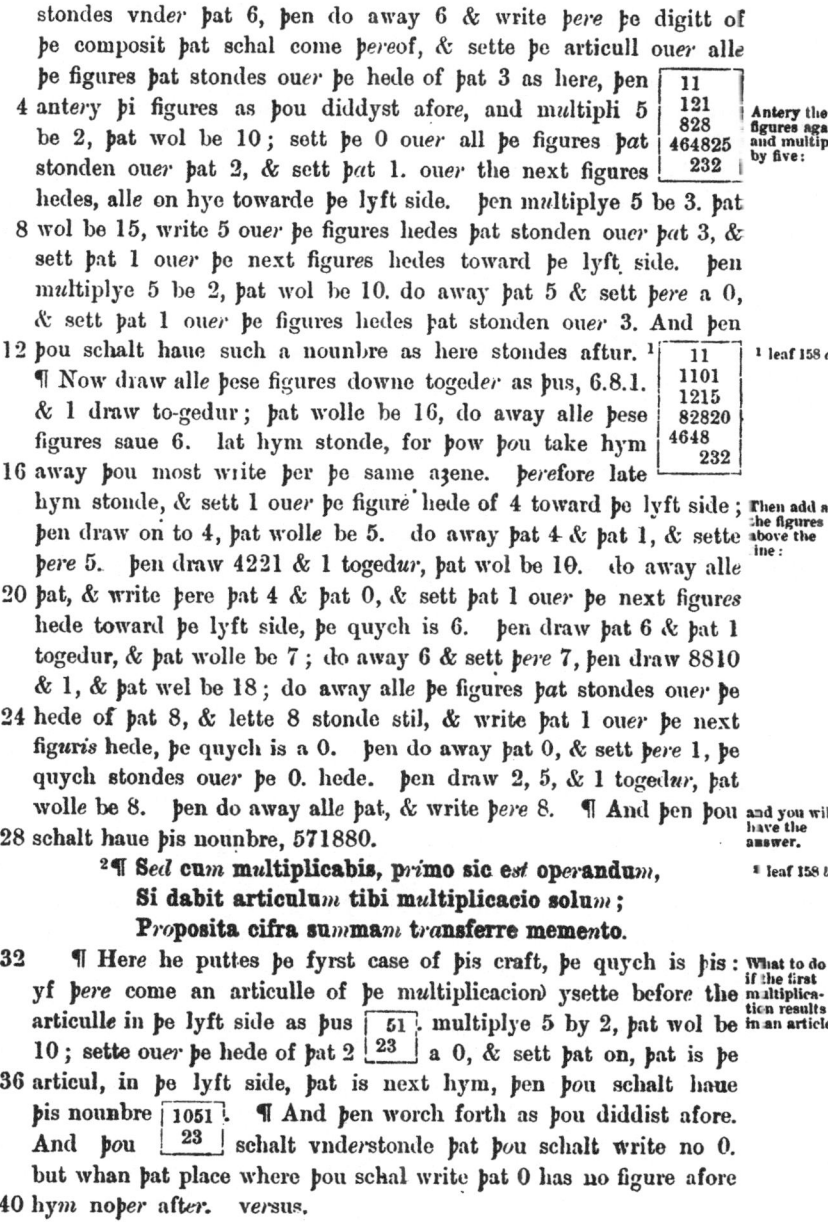

Antery the figures again, and multiply by five:

[1] leaf 158 *a*.

Then add all the figures above the line:

and you will have the answer.

[2] ¶ *Sed cum multiplicabis, primo sic est operandum,*
Si dabit articulum tibi multiplicacio solum;
Proposita cifra summam transferre memento.

[2] leaf 158 *b*.

¶ Here he puttes þe fyrst case of þis craft, þe quych is þis:
yf þere come an articulle of þe multiplicacion ysette before the
articulle in þe lyft side as þus multiplye 5 by 2, þat wol be
10 ; sette ouer þe hede of þat 2 a 0, & sett þat on, þat is þe
articul, in þe lyft side, þat is next hym, þen þou schalt haue
þis nounbre ¶ And þen worch forth as þou diddist afore.
And þou schalt vnderstonde þat þou schalt write no 0.
but whan þat place where þou schal write þat 0 has no figure afore
hym noþer aftur.   *versus.*

What to do if the first multiplication results in an article.

¶ **Si aut*em* digitus excreu*er*it articul*us*que.**
Articul*us*[1] sup*r*aposi*t*o digito salit vltra.

What to do if the result is a composite number.

¶ Here is þe secunde case, þe quych is þis: yf hit happe þat þere come a composyt, þou schalt write þe digitte oue*r* þe hede of þe neþer figure by þe quych þou multipliest þe hier figure; and sett þe articul*le* next hym toward þe lyft side, as þou diddyst afore, as þ*us*
$\boxed{\begin{array}{r}83\\83\end{array}}$. Multiply 8 by 8, þat wol be 64. Write þe 4 oue*r* 8, þat is to say, oue*r* þe hede of þe neþer 8; & set 6, þe quych [2]is an articul, next aft*er*. And þen þou schalt haue such a nounb*r*e as is here, $\boxed{\begin{array}{r}6483\\83\end{array}}$[3], And þen worch forth.

[leaf 159 a.]

¶ **Si digitus t*amen* ponas ips*um* sup*er* ipsam.**

What if it be a digit.

¶ Here is þe thryde case, þe quych is þis: yf hit happe þat of þi m*u*ltiplicacioun come a digit, þou schalt write þe digit oue*r* þe hede of þe neþer figure, by the quych þou m*u*ltipliest þe hier*e* figure, for þis nedes no Ensampul.

¶ **Subdita m*u*ltiplica non hanc que [incidit] illi**
Delet ea*m* penit*us* scribens quod p*r*ouenit inde.

The fourth case of the craft.

¶ Here is þe 4 case, þe quych is: yf hit be happe þat þe neþer figure schal multiplye þat figure, þe quych stondes oue*r* þat figures hede, þou schal do away þe hier figure & sett þere þat þat comys of þat m*u*ltiplicacion. As yf þere come of þat m*u*ltiplicacion an articuls þou schalt write þere þe hier figure stode a 0. ¶ And write þe articuls in þe lyft side, yf þat hit be a digit write þere a digit. yf þat h*i*t be a composit, write þe digit of þe composit. And þe articul in þe lyft side. al þis is ly3t y-now3t, þere-fore þer nedes no Ensampul.

¶ **Se*d* si multiplicat alia*m* ponas sup*er* ipsam**
Adiu*n*ges num*er*u*m* quem p*r*ebet duct*us* earu*m*.

[leaf 159 b.]
The fifth case of the craft.

¶ Here is þe 5 case, þe quych is þis: yf [4]þe neþer figure schul m*u*ltiplie þe hier, and þat hier figure is not recte oue*r* his hede. And þat neþer figure hase oþ*er* figures, or on figure oue*r* his hede by m*u*ltiplicacion, þat hase be afore, þou schalt write þat nounbre, þe quych comes of þat, oue*r* alle þe ylke figures hedes, as þus here:
$\boxed{\begin{array}{r}236\\234\end{array}}$ Multiply 2 by 2, þat wol be 4; set 4 oue*r* þe hede of þat 2. þen[5] m*u*ltiplies þe hier 2 by þe neþer 3, þat wol be 6. set oue*r* his hede 6, multiplie þe hier 2 by þe neþer 4, þat wol be 8. do away þe hier 2, þe quych stondes oue*r* þe hede of þe figure of 4,

---

[1] 'sed' deleted in MS.   [3] 6883 in MS.
[5] 'þen' overwritten on 'þat' marked for erasure.

and set þere 8. And þou schalt haue þis nounbre here ⌐46836⌐. And antery þi figures, þat is to say, set þi neþer 4 vnder þe ⌊234⌋ hier 3, and set þi 2 other figures nere hym, so þat þe neþer 2 stonde vndur 4 þe hier 6, þe quych 6 stondes in þe lyft side. And þat 3 þat stondes vndur 8, as þus aftur ȝe may se, ⌐46836⌐ Now worch forthermore, And multiplye þat hier 3 by 2, ⌊234⌋ þat wol be 6, set þat 6 þe quych stondes ouer þe hede of þat 2, And þen worch as I taȝt þe 8 afore.

    [1]¶ **Si supraposita cifra debet multiplicare**
       Prorsus eam deles & ibi scribi cifra debet.   [1] leaf 160 a.

  ¶ Here is þe 6 case, þe quych is þis: yf hit happe þat þe figure 12 by þe quych þou schal multiplye þe hier figure, þe quych stondes ryght ouer hym by a 0, þou schalt do away þat figure, þe quych ouer þat cifre hede. ¶ And write þere þat nounbre þat comes of þe multiplicacion as þus, 23. do away 2 and sett þere a 0. vnde 16 versus.   The sixth case of the craft.

    ¶ **Si cifra multiplicat aliam positam super ipsam**
       Sitque locus supra vacuus super hanc cifram fiet.

  ¶ Here is þe 7 case, þe quych is þis: yf a 0 schal multiply a 20 figure, þe quych stondes not recte ouer hym, And ouer þat 0 stonde no thyng, þou schalt write ouer þat 0 anoþer 0 as þus: ⌐24⌐ multiplye 2 be a 0, it wol be nothynge. write þere a 0 ouer þe ⌊03⌋ hede of þe neþer 0, And þen worch forth til þou come to þe ende.   The seventh case of the craft.

24   ¶ **Si supra[2] fuerit cifra semper est pretereunda.**

  ¶ Here is þe 8 case, þe quych is þis: yf þere be a 0 or mony cifers in þe hier rewe, þou schalt not multiplie hem, bot let hem stonde. And antery þe figures beneþe to þe next figure sygnificatyf 28 as þus: ⌐00032⌐. Ouer-lepe alle þese cifers & sett þat [3]neþer 2 þat stondes ⌊22⌋ toward þe ryght side, and sett hym vndur þe 3, and sett þe oþer nether 2 nere hym, so þat he stonde vndur þe thrydde 0, þe quych stondes next 3. And þan worch. vnde versus.   The eighth case of the craft.   [3] leaf 160 b.

32   ¶ **Si dubites, an sit bene multiplicacio facta,**
       Diuide totalem numerum per multiplicantem.

  ¶ Here he teches how þou schalt know wheþer þou hase wel I-do or no. And he says þat þou schalt deuide alle þe nounbre þat 36 comes of þe multiplicacion by þe neþer figures. And þen þou schalt haue þe same nounbur þat þou hadyst in þe begynnynge. but ȝet þou hast not þe craft of dyuision, but þou schalt haue hit afterwarde.   How to prove the multiplication.

    [2] 'Supra' inserted in MS. in place of 'cifra' marked for erasure.

¶ **Per numerum si vis numerum quoque multiplicare**
¶ **Tantum per normas subtiles absque figuris**
Has normas poteris per versus scire sequentes.

Mental multiplication.
¶ Here he teches þe to multiplie be þowȝt figures in þi mynde. 4
And þe sentence of þis verse is þis: yf þou wel multiplie on nounbre by anoþer in þi mynde, þou schal haue þereto rewles in þe verses þat schal come after.

¶ **Si tu per digitum digitum vis multiplicare** 8
Regula precedens dat qualiter est operandum.

Digit by digit is easy.
1 leaf 161 a.
¶ Here he teches a rewle as þou hast afore to multiplie a digit be anoþer, as yf þou wolde wete qwat is sex tymes 6. þou ¹schalt wete by þe rewle þat I taȝt þe before, yf þou haue mynde þerof. 12

¶ **Articulum si per reliquum reliquum vis multiplicare**
In proprium digitum debet vterque resolui.
¶ **Articulus digitos post se multiplicantes**
Ex digitus quociens retenerit multiplicari 16
Articuli faciunt tot centum multiplicati.

The first case of the craft.
¶ Here he teches þe furst rewle, þe quych is þis: yf þou wel multiplie an articul be anoþer, so þat both þe articuls bene with-Inne an hundreth, þus þou schalt do. take þe digit of bothe the 20

Article by article;
articuls, for euery articul hase a digit, þen multiplye þat on digit by þat oþer, and loke how mony vnytes ben in þe nounbre þat comes of þe multiplicacion of þe 2 digittes, & so mony hundrythes ben in þe nounbre þat schal come of þe multiplicacion of þe ylke 2 articuls 24

an example:
as þus. yf þou wold wete qwat is ten tymes ten. take þe digit of ten, þe quych is 1; take þe digit of þat oþer ten, þe quych is on.
¶ Also multiplie 1 be 1, as on tyme on þat is but 1. In on is but on vnite as þou wost welle, þerefore ten tymes ten is but a hun- 28

another example:
dryth. ¶ Also yf þou wold wete what is twenty tymes 30. take þe digit of twenty, þat is 2; & take þe digitt of thrytty, þat is 3. multiplie 3 be 2, þat is 6. Now in 6 ben 6 vnites, ¶ And so mony

² leaf 161 b.
hundrythes ben in 20 tymes 30², þerefore 20 tymes 30 is 6 hun- 32 dryth euen. loke & se. ¶ But yf it be so þat one articul be with-Inne an hundryth, or by-twene an hundryth and a thowsande, so þat it be not a þowsande fully. þen loke how mony vnytes ben in þe nounbur þat comys of þe multiplicacion ³And so mony tymes³ 36 of 2 digittes of ylke articuls, so mony thowsant ben in þe nounbre, the qwych comes of þe multiplicacion. And so mony tymes ten thowsand schal be in þe nounbre þat comes of þe multiplicacion of

³⁻³ Marked for erasure in MS.

2 articuls, as yf þou wold wete qwat is 4 hundryth tymes [two hundryth]. Multiply 4 be 2,[1] þat wol be 8. in 8 ben 8 vnites.

¶ And so mony tymes ten thousand be in 4 hundryth tymes 4 [2][1] hundryth, þat is 80 thousand. Take hede, I schall telle þe a generalle rewle whan þou hast 2 articuls, And þou wold wete qwat comes of þe multiplicacioṅ of hem 2. multiplie þe digit of þat on articuls, and kepe þat nounbre, þen loke how mony cifers schuld go 8 before þat on articuls, and he were write. Als mony cifers schuld go before þat other, & he were write of cifers. And haue alle þe ylke cifers togedur in þi mynde, [2]a-rowe ychoṅ aftur other, and in þe last plase set þe nounbre þat comes of þe multiplicacioṅ of þe 12 2 digittes. And loke in þi mynde in what place he stondes, where in þe secunde, or in þe thryd, or in þe 4, or where ellis, and loke qwat þe figures by-token in þat place; & so mych is þe nounbre þat comes of þe 2 articuls y-multiplied to-gedur as þus: yf þou wold 16 wete what is 20 thousand tymes 3 þowsande. multiply þe digit of þat articulle þe quych is 2 by þe digitte of þat oþer articul þe quych is 3, þat wol be 6. þen loke how mony cifers schal go to 20 thousand as hit schuld be write in a tabul. certainly 4 cifers schuld go to 20 20 þowsant. ffor þis figure 2 in þe fyrst place betokenes twene.

¶ In þe secunde place hit betokenes twenty. ¶ In þe 3. place hit betokenes 2 hundryth. .¶. In þe 4 place 2 thousant. ¶ In þe 5 place hit betokenes twenty þousant. þerefore he most haue 4 cifers 24 a-fore hym þat he may stonde in þe 5 place. kepe þese 4 cifers in thy mynde, þen loke how mony cifers goṅ to 3 thousant. Certayn to 3 thousante [3]goṅ 3 cifers afore. Now cast ylke 4 cifers þat schuld go to twenty thousant, And thes 3 cifers þat schuld go 28 afore 3 thousant, & sette hem in rewe ychoṅ after oþer in þi mynde, as þai schuld stonde in a tabulle. And þen schal þou haue 7 cifers; þen sett þat 6 þe quych comes of þe multiplicacioṅ of þe 2 digittes aftur þe ylke cifers in þe 8 place as yf þat hit stode in a 32 tabul. And loke qwat a figure of 6 schuld betoken in þe 8 place. yf hit were in a tabul & so mych it is. & yf þat figure of 6 stonde in þe fyrst place he schuld betoken but 6. ¶ In þe 2 place he schuld betoken sexty. ¶ In the 3 place he schuld betokeṅ sex hundryth. 36 ¶ In þe 4 place sex thousant. ¶ In þe 5 place sexty þowsant. ¶ In þe sext place sex hundryth þowsant. ¶ In þe 7 place sex þowsant thousantes. ¶ In þe 8 place sexty þowsant thousantes. þerfore sett 6 in octauo loco, And he schal betoken sexty þowsant

Mental multi-plication.

Another example.

[2] leaf 162 a.

Another example.

Notation.

[3] leaf 162 b.

Notation again.

---
[1] 4 in MS.

The Craft of Multiplication.

Mental multiplication.

thousantes. And so mych is twenty þowsant tymes 3 thousant, ¶ And þis rewle is generalle for alle maner of articuls, Whethir þai be hundryth or þowsant; but þou most know well þe craft of þe wryrchynge in þe tabulle [1]or þou know to do þus in þi mynde 4 aftur þis rewle. Thou most þat þis rewle holdyþe note but where þere ben 2 articuls and no mo of þe quych ayther of hem hase but on figure significatyf. As twenty tymes 3 thousant or 3 hundryth, and such oþur. 8

1 leaf 163 a.

> ¶ Articulum digito si multiplicare oportet
> Articuli digit[i sumi quo multiplicate]
> Debemus reliquum quod multiplicatur ab illis
> Per reliquo decuplum sic summam latere nequibit. 12

The third case of the craft;

¶ Here he puttes þe thryde rewle, þe quych is þis. yf þou wel multiply in þi mynde, And þe Articul be a digitte, þou schalt loke þat þe digitt be with-Inne an hundryth, þen þou schalt multiply the digitt of þe Articulle by þe oþer digitte. And euery vnite in þe 16 nounbre þat schalle come þere-of schal betoken ten. As þus: yf þat þou wold wete qwat is twyes 40. multiplie þe digitte of 40, þe quych is 4, by þe oþer diget, þe quych is 2. And þat wolle be 8. And in þe nombre of 8 ben 8 vnites, & euery of þe ylke vnites 20 schuld stonde for 10. þere-fore þere schal be 8 tymes 10, þat wol be 4 score. And so mony is twyes 40. ¶ If þe articul be a hundryth or be 2 hundryth And a þowsant, so þat hit be notte a thousant, [2]worch as þou dyddyst afore, saue þou schalt rekene euery 24 vnite for a hundryth.

an example.

2 leaf 163 b.

> ¶ In numerum mixtum digitum si ducere cures
> Articulus mixti sumatur deinde resoluas
> In digitum post fac respectu de digitis 28
> Articulusque docet excrescens in diriuando
> In digitum mixti post ducas multiplicantem
> ¶ De digitis vt norma [3][docet] de [hunc]
> Multiplica simul et sic postea summa patebit. 32

The fourth case of the craft:

Here he puttes þe 4 rewle, þe quych is þis: yf þou multipliy on composit be a digit as 6 tymes 24, [4]þen take þe diget of þat composit, & multiply þat digitt by þat oþer diget, and kepe þe nombur þat comes þere-of. þen take þe digit of þat composit, & multiply þat 36 digit by anoþer diget, by þe quych þou hast multiplyed þe diget of þe articul, and loke qwat comes þere-of. þen take þou þat nounbur, & cast hit to þat other nounbur þat þou secheste as þus yf þou wel

Composite by digit.

[3] docet. decet MS. [4] '4 times 4' in MS.

wete qwat comes of 6 tymes 4 & twenty.   multiply þat articulle of Mental mul-
þe composit by þe digit, þe quych is 6, as yn þe thryd rewle þou   tiplication.
was tauȝt, And þat schal be 6 score.   þen multiply þe diget of þe
4 composit, ¹þe quych is 4, and multiply þat by þat other diget, þe ¹ leaf 164 a.
quych is 6, as þou wast tauȝt in þe first rewle, yf þou haue mynde
þerof, & þat wol be 4 & twenty. ⁻cast all ylke nounburs to-gedir,
& hit schal be 144.   And so mych is 6 tymes 4 & twenty.

8   ¶ **Ductus in articulum numerus si compositus sit**
      Articulum purum comites articulum quoque
      Mixti pro digitis post fiat [et articulus vt]
      Norma iubet [retinendo quod extra dicta ab illis]
12   **Articuli digitum post tu mixtum digitum duc**
      Regula de digitis nec precipit articulusque
      Ex quibus excrescens summe tu iunge priori
      Sic manifesta cito fiet tibi summa petita.

16   ¶ Here he puttes þe 5 rewle, þe quych is þis: yf þou wel The fifth case
multiply an Articul be a composit, multiplie þat Articul by þe   of the craft:
articul of þe composit, and worch as þou wos tauȝt in þe secunde
rewle, of þe quych rewle þe verse begynnes þus. ¶ **Articulum si** Article by
20 **per Relicum vis multiplicare.** þen multiply þe diget of þe composit Composite.
by þat oþir articul aftir þe doctrine of þe 3 rewle.   take þerof gode
hede, I pray þe as þus. Yf þou wel wete what is 24 tymes ten.
Multiplie ten by 20, þat wel be 2 hundryth. þen multiply þe diget An example.
24 of þe 10, þe quych is 1, by þe diget of þe composit, þe quych is 4,
& þat ²wol be 4. þen reken euery vnite þat is in 4 for 10, & þat ² leaf 164 b.
schal be 40. Cast 40 to 2 hundryth, & þat wol be 2 hundryth & 40.
And so mych is 24 tymes ten.

28   ¶ **Compositum numerum mixto si[c] multiplicabis**
      Vndecies tredecim sic est ex hiis operandum
      In reliquum primum demum duc post in eundem
      Vnum post denum duc in tria deinde per vnum
32   **Multiplicesque demum intra omnia multiplicata**
      In summa decies quam si fuerit tibi doces
      Multiplicandorum de normis sufficiunt hec.

¶ Here he puttes þe 6 rewle, & þe last of alle multiplicacion, The sixth case
36 þe quych is þis: yf þou wel multiplye a composit by a-noþer com- of the craft:
posit, þou schalt do þus. multiplie þat on composit, qwych þou welt Composite by
of the twene, by þe articul of þe toþer composit, as þou were tauȝt in   Composite.
þe 5 rewle, þen multiplie þat same composit, þe quych þou hast
40 multiplied by þe oþer articul, by þe digit of þe oþer composit, as

Mental multiplication. An example

þou was tauȝt in þe 4 rewle. As þus, yf þou wold wete what is 11 tymes 13, as þou was tauȝt in þe 5 rewle, & þat schal be an hundryth & ten, afterwarde multiply þat same composit þat þou hast multiplied, þe quych is a .11. And multiplye hit be þe digit of þe oþer composit, þe quych is 3, for 3 is þe digit of 13, And þat wel be 30. þen take þe digit of þat composit, þe quych composit þou multiplied by þe digit of þat oþer composit, [1]þe quych is a 11. ¶ Also of þe quych 11 on is þe digit. multiplie þat digitt by þe digett of þat other composit, þe quych diget is 3, as þou was tauȝt in þe first rewle in þe begynnynge of þis craft. þe quych rewle begynnes "In digitum cures." And of alle þe multiplicacion of þe 2 digitt comys thre, for onys 3 is but 3. Now cast alle þese nounbers togedur, the quych is þis, a hundryth & ten & 30 & 3. And al þat wel be 143. Write 3 first in þe ryght side. And cast 10 to 30, þat wol be 40. set 40 next aftur towarde þe lyft side, And set aftur a hundryth as here an Ensampulle, 143.

[1] leaf 165 a.

of the sixth case of the craft.

4

8

12

16

(Cetera desunt.)

The Art of Nombryng.

A TRANSLATION OF
John of Holywood's De Arte Numerandi.

[*Ashmole MS. 396, fol.* 48.]

Boys seying in the begynnyng of his Arsemetrike :—Alle thynges that bene fro the first begynnyng of thynges have procedede, and come forthe, And by resoun of nombre ben formede; And in wise as they bene, So owethe
4 they to be knowene; wherfor in vniuersalle knowlechyng of thynges the Art of nombrynge is best, and most operatyfe. Fol. 48.

Therfore sithen the science of the whiche at this tyme we intendene to write of standithe alle and about nombre:
8 ffirst we most se, what is the propre name therofe, and fro whens the name come: Afterwarde what is nombre, And how manye spices of nombre ther ben. The name is clepede Algorisme,
12 hade out of Algore, other of Algos, in grewe, That is clepide in englisshe art other craft, And of Rithmus that is callede nombre. So algorisme is clepede the art of nombryng, other it is had ofe en or in, and gogos that is introducciown, and Rithmus nombre, that is
16 to say Interducciown of nombre. And thirdly it is hade of the name of a kyng that is clepede Algo and Rythmus; So callede Algorismus. Sothely .2. manere of nombres ben notifiede; Formalle,[1] as nombre is vnitees gadrede to-gedres; Materialle,[2] as
20 nombre is a colleccioun of vnitees. Other nombre is a multitude hade out of vnitees, vnitee is that thynge wher-by euery thynge is callede oone, other o thynge. Of nombres, that one is clepede digitalle, that othere Article, Another a nombre componede oþer
24 myxt. Another digitalle is a nombre with-in .10.; Article is þat nombre that may be dyvydede in .10. parties egally, And that there

The name of the art.

Derivation of Algorism.

Another.

Another.

Kinds of numbers.

---
[1] MS. Materialle.   [2] MS. Formalle.

34   Chapter I. Numeration.

leve no residue; Componede or medlede is that nombre that is
come of a digite and of an article. And vndrestande wele that alle
nombres betwix .2. articles next is a nombre componede. Of this
art bene .9. spices, that is forto sey, numeracioun, addicioun, Subtraccioun, Mediacioun, Duplacioun, Multipliacioun, Dyvysioun, Progressioun, And of Rootes the extraccioun, and that may be hade in
.2. maners, that is to sey in nombres quadrat, and in cubices:
Amonge the whiche, ffirst of Numeracioun, and afterwarde of þe
oþers by ordure, y entende to write.

^{margin: The 9 rules of the Art.}
^{margin line numbers: 4, 8}

^{margin: 1 Fol. 48 b.}

[1]**For-sothe numeracioun is of euery numbre by competent
figures an artificialle representacioun.**

^{margin: Figures, differences, places, and limits.}

Sothly figure, difference, places, and lynes supposen o thyng
other the same, But they ben sette here for dyuers resons.
ffigure is clepede for protraccioun of figuracioun; Difference is
callede for therby is shewede euery figure, how it hathe difference
fro the figures before them: place by cause of space, where-in me
writethe: lynees, for that is ordeynede for the presentacioun of
euery figure. And vnderstonde that ther ben .9. lymytes of
figures that representen the .9. digites that ben these. 0. 9. 8. 7. 6.
5. 4. 3. 2. 1. The .10. is clepede theta, or a cercle, other a cifre,
other a figure of nought for nought it signyfiethe. Nathelesse she
holdyng that place givethe others for to signyfie; for withe-out cifre
or cifres a pure article may not be writte. And sithen that by
these .9. figures significatifes Ioynede with cifre or with cifres alle
nombres ben and may be representede, It was, nether is, no nede to
fynde any more figures. And note wele that euery digite shalle be
writte with oo figure allone to it aproprede. And alle articles by
a cifre, ffor euery article is namede for oone of the digitis as .10. of
1.. 20. of. 2. and so of the others, &c. And alle nombres digitalle
owen to be sette in the first difference: Alle articles in the seconde.
Also alle nombres fro .10. til an .100. [which] is excludede, with .2.
figures mvst be writte; And yf it be an article, by a cifre first put,
and the figure y-writte towarde the lift honde, that signifiethe the
digit of the whiche the article is namede; And yf it be a nombre
componede, ffirst write the digit that is a part of that componede,
and write to the lift side the article as it is seide be-fore. Alle
nombre that is fro an hundrede tille a thousande excludede, owithe
to be writ by .3. figures; and alle nombre that is fro a thousande

^{margin: The 9 figures.}
^{margin: The cipher.}
^{margin: The numeration}
^{margin: of digits,}
^{margin: of articles,}
^{margin: of composites.}
^{margin line numbers: 12, 16, 20, 24, 28, 32, 36}

Chapter II. Addition.

til .x. Mł. mvst be writ by .4. figures; And so forthe. And vnderstonde wele that every figure sette in the first place signyfiethe his digit; In the seconde place .10. tymes his digit; In the .3. place an 4 hundrede so moche; In the .4. place a thousande so moche; In the .5. place .x. thousande so moche; In the .6. place an hundrede thousande so moche; In the .7. place a thousande thousande. And so infynytly mvltiplying by ¹these .3. 10, 100, 1000. And vnder-8 stande wele that competently me may sette vpon figure in the place of a thousande, a prike to shewe how many thousande the last figure shalle represent. We writene in this art to the lift side-warde, as arabiene writene, that weren fynders of this science, othere for this 12 resoun, that for to kepe a custumable ordre in redyng, Sette we alle-wey the more nombre before.

<small>The value due to position.</small>

<small>¹ Fol. 49.</small>

<small>Numbers are written from right to left.</small>

Addicioun is of nombre other of nombres vnto nombre or to nombres aggregacioun, that me may see that that is come 16 therof as excressent. In addicioun, 2. ordres of figures and .2. nombres ben necessary, that is to sey, a nombre to be addede and the nombre wherto the addicioun sholde be made to. The nombre to be addede is that þat sholde be addede therto, and shalle 20 be vnderwriten; the nombre vnto the whiche addicioun shalle be made to is that nombre that resceyuethe the addicion of þat other, and shalle be writen above; and it is convenient that the lesse nombre be vnderwrit, and the more addede, than the contrary. 24 But whether it happe one other other, the same comythe of, Therfor, yf þow wilt adde nombre to nombre, write the nombre wherto the addicioun shalle be made in the omest ordre by his differences, so that the first of the lower ordre be vndre the first 28 of the omyst ordre, and so of others. That done, adde the first of the lower ordre to the first of the omyst ordre. And of suche addicioun, other þere growith therof a digit, An article, other a composede. If it be digitus, In the place of the omyst shalt thow 32 write the digit excrescyng, as thus :—

<small>Definition.</small>

<small>How the numbers should be written.</small>

<small>The method of working.</small>

<small>Begin at the right.</small>

The resultant	2
To whom it shal be addede	1
The nombre to be addede	1

If the article; in the place of the omyst put a-way by a cifre. writte, and the digit transferrede, of þe 36 whiche the article toke his name, towarde the lift side, and be it addede to the next figure folowyng, yf ther be any figure folowyng; or no, and yf it be not, leve it [in the] voide, as thus :—

<small>The Sum is a digit,</small>

Chapter III. Subtraction.

or an article,

The resultant	10
To whom it shalle be addede	7
The nombre to be addede	3

Resultans	2	7	8	2	7
Cui debet addi	1	0	0	8	4
Numerus addendus	1	7	7	4	3

And yf it happe that the figure folowyng wherto the addicioun shalle be made by [the cifre of] an article, it sette a-side; In his place write the [1][digit of the] Article as thus:—

[1] Fol. 49 b.

The resultant	17
To whom it shalle be addede	10
The nombre to be addede	7

4

And yf it happe that a figure of .9. by the figure that me mvst adde [one] to, In the place of that 9. put a cifre *and* write þe article towarde þe lift honde as bifore, and thus:—

The resultant	10
To whom it shalle be addede	9
The nombre to be addede	1

8

or a composite.

And yf[2] [therefrom grow a] nombre componed,[3] [in the place of the nombre] put a-way[4] [let] the digit [be][5] writ þat is part of þat composide, and þan put to þe lift side the article as before, and þus:—

The resultant	12
To whom it shalle be addede	8
The nombre to be addede	4

12

The translator's note.

This done, adde the seconde to the seconde, and write above oþer as before. Note wele þat in addicions and in alle spices folowyng, whan he seithe one the other shalle be writen aboue, and me most vse euer figure, as that euery figure were sette by halfe, and by hym-selfe.

16

Definition of Subtraction.

Subtraccioun is of .2. proposede nombres, the fyndyng of the excesse of the more to the lasse: Other subtraccioun is ablacioun of o nombre fro a-nother, that me may see a some left. The lasse of the more, or even of even, may be withdraw; The more fro the lesse may neuer be. And sothly that nombre is more that hathe more figures, So that the last be signyficatifes: And yf ther ben as many in that one as in that other, me most deme it by the last, other by the next last. More-ouer in withdrawyng .2. nombres ben necessary; A nombre to be withdraw, And a nombre that me shalle with-draw of. The nombre to be with-draw shalle be writ in the lower ordre by his differences; The

20

How it may be done.

24

What is required.

28

---

[2] 'the' in MS. [3] 'be' in MS. [4] 'and' in MS.
[5] 'is' in MS.

Chapter III. Subtraction.

nombre fro the whiche me shalle withe-draw in the omyst ordre, so that the first be vnder the first, the seconde vnder the seconde, And so of alle others. Withe-draw therfor the first of the lowere ordre fro the first of the ordre above his hede, and that wolle be other more or lesse, oþer egalle. yf it be egalle or even the figure sette beside, put in his place a cifre. And yf it be more put away þerfro als many of vnitees the lower figure conteynethe, and writ the residue as thus

Write the greater number above.
Subtract the first figure if possible.

The remanent	20
Wherof me shalle witℏdraw	22
The nombre to be witℏdraw	2

The remanent	2	2
Wherof me shalle witℏ-draw	2	8
þe nombre to be witℏdraw		6

Remanens	2	2	1	8	2	9	9	9	8
A quo sit subtraccio	8	7	2	4	3	0	0	0	4
Numerus subtrahendus	6	5	2	[6]	.	.	.	.	6

And yf it be lesse, by-cause the more may not be with-draw ther-fro, borow an vnyte of the next figure that is worthe 10. Of that .10. and of the figure that ye wolde have witℏ-draw fro be-fore to-gedre Ioynede, witℏ-draw þe figure be-nethe, and put the residue in the place of the figure put a-side as þus:—
And yf the figure wherof me shal borow the vnyte be one,

[1] Fol. 50.
If it is not possible borrow ten,
and then subtract.

The remanent	1	8
Wherof me shalle witℏ-draw	2	4
The nombre to be witℏ-draw	0	6

If the second figure is one.

put it a-side, and write a cifre in the place þerof, lest the figures folowing faile of thaire nombre, and þan worche as it shewitℏ in this figure here:—
And yf the vnyte wherof me shal borow be a cifre, go ferther to the figure signy-

The remanent	3	0	9[2]
Wherof me shal witℏ-draw	3	1	2
The nombre to be witℏ-draw	.	.	3

If the second figure is a cipher.

ficatife, and ther borow one, and retournyng bake, in the place of euery cifre þat ye passide ouer, sette figures of .9. as here it is specifiede:—
And whan me comethe to the nombre wherof me intendithe, there re-

The remenaunt	2	9	9	9	9
Wherof me shalle witℏ draw	3	0	0	0	3
The nombre to be witℏ-draw					4

maynethe alle-wayes .10. ffor þe whiche .10. &c. The reson why þat for euery cifre left behynde me setteth figures ther of .9. this it is:—If fro the .3. place me borowede an vnyte, that vnyte by respect of the figure that he came fro representith an .C., In the

A justification of the rule given.

---

[2] 6 in MS.   [3] 0 in MS.

Chapter IV. Mediation.

place of that cifre [passed over] is left .9., [which is worth ninety], and yit it remaynethe as .10., And the same resone wolde be yf me hade borowede an vnyte fro the .4., .5., .6., place, or ony other so vpwarde. This done, withdraw the seconde of the lower ordre fro the figure above his hede of þe omyst ordre, and wirche as before. *And note wele that in addicion or in subtraccioun me may wele fro the lift side begynne and ryn to the right side, But it wol be more profitabler to be do, as it is taught.* And yf thow wilt prove yf thow have do wele or no, The figures that thow hast withdraw, adde them ayene to the omyst figures, and they wolle accorde with the first that thow haddest yf thow have labored wele; and in like wise in addicioun, whan thow hast addede alle thy figures, withdraw them that thow first ¹addest, and the same wolle retourne. The subtraccioun is none other but a prouffe of the addicioun, and the contrarye in like wise.

<small>Why it is better to work from right to left.</small>

<small>How to prove subtraction,</small>

<small>and addition.</small>

<small>¹ Fol. 50 b.</small>

Mediacioun is the fyndyng of the halfyng of euery nombre, that it may be seyne what and how moche is euery halfe. In halfyng ay oo order of figures and oo nombre is necessary, that is to sey the nombre to be halfede. Therfor yf thow wilt half any nombre, write that nombre by his differences, and begynne at the right, that is to sey, fro the first figure to the right side, so that it be signyficatife other represent vnyte or eny other digitalle nombre. If it be vnyte write in his place a cifre for the figures folowyng, [lest they signify less], and write that vnyte without in the table, other resolue it in .60. mynvtes and sette a-side half of tho minutes so, and reserve the remenaunt without in the table, as thus .30. ; other sette without thus .ā : that kepethe none ordre of place, Nathelesse it hathe signyficacioun. And yf the other figure signyfie any other digital nombre fro vnyte forthe, oþer the nombre is ode or evene. If it be even, write this half in this wise :—
And if it be odde, Take the next even vndre hym conteynede, and put his half in the place of that odde, and of þe vnyte that remaynethe to be halfede do thus :—
This done, the seconde is to be halfede, yf it be a cifre put it be-side, and yf it be significatife, other it is even or ode : If it be even, write in the place of þe nombres wipede out the halfe ; yf it be ode, take the next even vnder it contenythe, and in the place of the Impar sette a-side put half of the even : The

<small>Definition of mediation.</small>

<small>Where to begin.</small>

<small>If the first figure is unity.</small>

<small>What to do if it is not unity.</small>

<small>Then halve the second figure.</small>

| Halfede | 2 | 2 |
| to be halfede | 4 | 4 |

| halfede | 2 | 3 | [di] |
| To be halfede | 4 | 7 | |

Chapter V. Duplation.

vnyte that remaynethe to be halfede, respect hade to them before, is worthe .10. Dyvide that .10. in .2., 5. is, and sette a-side that one, and adde that other to the next figure
4 precedent as here:—
And yf þe addicioun sholde be made to a cifre, sette it a-side, and write in his place .5. And vnder this fourme me shalle write and worche,
8 tille the totalle nombre be halfede.

If it is odd, add 5 to the figure before.

Halfede			
to be halfede			

doublede	2	6	8	9	0	10	17	4
to be doublede	1	3	4	4	5	5	8	7

Duplicacioun is agregacion of nombre [to itself] þat me may se the nombre growen. In doublynge ay is but one ordre of
12 figures necessarie. And me most be-gynne with the lift side, other of the more figure, And after the nombre of the more figure representithe. ¹In the other .3. before we begynne alle way fro the right side and fro the lasse nombre, In this spice and in alle
16 other folowyng we wolle begynne fro the lift side, ffor and me bigon the double fro the first, omwhiie me myght double oo thynge twyes. And how be it that me myght double fro the right, that wolde be harder in techyng and in workyng. Therfor yf thow
20 wolt double any nombre, write that nombre by his differences, and double the last. And of that doublyng other growithe a nombre digital, article, or componede. [If it be a digit, write it in the place of the first digit.] If it be article, write in his place a cifre
24 and transferre the article towarde the lift, as thus:—
And yf the nombre be componede, write a digital that is part of his composicioun, and sette the article to the
28 lift hande, as thus:—
That done, me most double the last save one, and what growethe þerof me most worche as before. And yf a cifre be, touche it not. But yf any nombre
32 shalle be addede to the cifre, in þe place of þe figure wipede out me most write the nombre to be addede, as thus:—
In the same wise me shalle wirche of
36 alle others. And this probacioun: If thow truly double the halfis, and truly half the doubles, the same nombre and figure shalle mete, suche as thow labourede vpone first, And of the
40 contrarie.

Definition of Duplation.

¹ Fol. 51.

Where to begin.

Why.

What to do with the result.

double	10
to be doublede	5

doublede	16
to be doublede	8

doublede	6	0	6
to be doublede	3	0	3

How to prove your answer.

Doublede	6	1	8
to be doublede	3	0	9

Chapter VI. Multiplication.

Definition of Multiplication.

Multiplicacioun of nombre by hym-self other by a-nother, with praposide .2. nombres, [is] the fyndyng of the thirde, That so oft conteynethe that other, as ther ben vnytes in the oþer. In multiplicacioun .2. nombres pryncipally ben necessary, that is to sey, the nombre multiplying and the nombre to be **Multiplier.** multipliede, as here;—twies fyve. [The number multiplying] is designede aduerbially. The nombre to be multipliede resceyvethe **Multiplicand.** a nominalle appellacioun, as twies .5. 5. is the nombre multipliede, and twies is the nombre to be multipliede.

Resultans	[1]	1	0	1	3	2	6	6	8	0	0	8
Multiplicandus	.	.	5	.	.	4	.	3	4	0	0	4
Multiplicans	.	2	2	.	3	3	2	2	2	.	.	.

Product.
[2] Fol. 51 b.

Also me may thervpone to assigne the. 3. nombre, the whiche is [2]clepede product or provenient, of takyng out of one fro another: as twyes .5 is .10., 5. the nombre to be multipliede, and .2. the multipliant, and. 10. as before is come therof. And vnderstonde wele, that of the multipliant may be made the nombre to be multipliede, and of the contrarie, remaynyng euer the same some, and her-ofe comethe the comen speche, that seithe all nombre is convertede by Multiplying in hym-selfe.

1	2	3	4	5	6	7	8	9	10
2	4	6	8	10	10[3]	14	16	18	20
3	6	9	12	15	18	21	24	27	30
4	8	12	16	20	24	28	32	36	40
5	10	15	20	25	30	35	40	45	50
6	12	18	24	30	36	42	48	56	60
7	14	21	28	35	42	49	56	63	70
8	16	24	32	40	48	56	64	72	80
9	18	27	36	45	54	63	72	81	90
10	20	30	40	50	60	70	80	90	100

There are 6 rules of Multiplication.
And ther ben .6 rules of Multiplicacioun; ffirst, yf a digit multiplie a

(1) Digit by digit.
digit, considre how many of vnytees ben betwix the digit by multi-plying and his .10. bethe to-gedre accomptede, and so oft with-draw the digit multiplying, vnder the article of his denominacioun. Example of grace. If thow wolt wete how moche is .4. tymes .8., [4]se how many vnytees ben betwix .8.[5] and .10. to-geder rekenede, and it shewith that .2. : withdraw ther-for the quaternary, of the article of his denominacion twies, of .40., And ther remaynethe **See the table above.** .32., that is, to some of alle the multiplicacioun. Wher-vpon for more evidence and declaracion the seide table is made. Whan a **(2) Digit by article.** digit multipliethe an article, thow most bryng the digit into þe digit, of þe whiche the article [has][6] his name, and euery vnyte

---

[1] 2 in MS. [3] sic. [4] 'And' inserted in MS.
[5] '4 the' inserted in MS. [6] 'to' in MS.

shalle stonde for .10., and euery article an .100. Whan the digit (3) Composite by digit.
multipliethe a nombre componede, þou most bryng the digit into
aiþer part of the nombre componede, so þat digit be had into digit
4 by the first rule, into an article by þe seconde rule; and afterwarde
Ioyne the produccioun, and þere wol be the some totalle.

Resultans	1	2	6		7	3	6		1	2	0		1	2	0	8
Multiplicandus			2			3	2				6					4
Multiplicans		6	3		2	3				2	0			3	0	2

Whan an article multipliethe an article, the digit wherof he is (4) Article by article.
namede is to be brought Into the digit wherof the oþer is namede,
8 and euery vnyte wol be worthe [1]an .100., and euery article. a [1] Fol. 52.
.1000. Whan an article multipliethe a nombre componede, thow (5) Composite by article.
most bryng the digit of the article into aither part of the nombre
componede; and Ioyne the produccioun, and euery article wol be
12 worthe .100., and euery vnyte .10., and so wolle the some be
opene. Whan a nombre componede multipliethe a nombre com- (6) Composite by composite.
ponede, euery part of the nombre multiplying is to be hade into
euery part of the nombre to be multipliede, and so shalle the digit
16 be hade twies, onys in the digit, that other in the article. The
article also twies, ones in the digit, that other in the article. Therfor yf thow wilt any nombre by hym-self other by any other
multiplie, write the nombre to be multipliede in the ouer ordre by
20 his differences, The nombre multiplying in the lower ordre by his How to set down your numbers.
differences, so that the first of the lower ordre be vnder the last of
the ouer ordre. This done, of the multiplying, the last is to be
hade into the last of the nombre to be multipliede. Wherof than
24 wolle grow a digit, an article, other a nombre componede. If it be If the result is a digit,
a digit, even above the figure multiplying is hede write his digit
that come of, as it apperethe here:—

The resultant	6
To be multipliede	3
Þe nombre multipliyng	2

And yf an article had be writ ouer the figure multiplying his hede, an article,
28 put a cifre þer and transferre the article towarde the lift hande, as
thus:—

The resultant	1	0
to be multipliede		5
þe nombre multipliyng		2

And yf a nombre componede be writ ouer the figure multyplying is or a compo-
hede, write the digit in the nombre componede is place, and sette site.
32 the article to the lift hande, as thus:—

Chapter VI. Multiplication.

Multiply next by the last but one, and so on.

The resultant	1	2
To be multipliede		4
the nombre multiplying		3

This done, me must bryng the last save one of the multiplyng into the last of þe nombre to be multipliede, and se what comythe therof as before, and so do with alle, tille me come to the first of the nombre multiplying, that must be brought into the last of the nombre to be multipliede, wherof growithe oþer a digit, an article,

[1] Fol. 52 b.

[1]other a nombre componede. If it be a digit, In the place of the ouerer, sette a-side, as here:

Resultant	6	6
to be multipliede		3
the nombre multiplyng	2	2

If an article happe, there put a cifre in his place, and put hym to the lift hande, as here:

The resultant	1	1	0
to be multipliede			5
þe nombre multiplying		2	2

If it be a nombre componede, in the place of the ouerer sette a-side, write a digit that[2] is a part of the componede, and sette on the left honde the article, as here:

The resultant	1	3[3]	2
to be multipliede			4
þe nombre multipliant		3	3

Then antery the multiplier one place.

That done, sette forwarde the figures of the nombre multiplying by oo difference, so that the first of the multipliant be vnder the last save one of the nombre to be multipliede, the other by o place sette forwarde. Than me shalle brynge the last of the multipliant in hym to be multipliede, vnder the whiche is the first multipliant.

Work as before.

And than wolle growe oþer a digit, an article, or a componede nombre. If it be a digit, adde hym even above his hede; If it be an article, transferre hym to the lift side; And if it be a nombre componede, adde a digit to the figure above his hede, and sette to the lift hande the article. And alle-wayes euery figure of the nombre multipliant is to be brought to the last save one nombre to be multipliede, til me come to the first of the multipliant, where me shalle wirche as it is seide before of the first, and afterwarde to put forwarde the figures by o difference and one tille they alle be multipliede.

How to deal with ciphers.

And yf it happe that the first figure of þe multipliant be a cifre, and boue it is sette the figure signyficatife, write a cifre in the place of the figure sette a-side, as thus, etc.:

The resultant	1	2	0
to be multipliede			6
the multipliant		2	0

4
8
12
16
20
24
28
32

---

[2] 'that' repeated in MS.   [3] '1' in MS.

Chapter VII. Division.

And yf a cifre happe in the lower order be-twix the first and the last, and even above be sette the figure signyficatif, leve it vn-touchede, as here :— *How to deal with ciphers.*

And yf the space above sette be voide, in that place write thow a cifre.

The resultant		2	2	6	4	4
To be multipliede				2	2	2
The multipliant		1	0	2		

And yf the cifre happe betwix þe first and the last to be multipliede, me most sette forwarde the ordre of the figures by thaire differences, for oft of ducciown of figures in cifres nought is the resultant, as here,[1] wherof [1] Fol. 53.
it is evident and open, yf that the first figure of the nombre be to be multipliede be a cifre, vndir it shalle be none sette as here :—

Resultant		8	0	0	8	
to be multipliede		4	0	0	4	
the multipliant		2	.	.	.	.

Vnder [stand] also that in multiplica-ciown, divisiown, and of rootis the ex-tracciown, competently me may leve a mydel space betwix .2. ordres of figures, that me may write there what is come of addyng other withe-drawyng, lest any thynge sholde be ouer-hippede and sette out of mynde. *Leave room between the rows of figures.*

Resultant		3	2	0[1]
To be multipliede			8	0
The multipliant				4

For to dyvyde oo nombre by a-nother, it is of .2. nombres pro-posede, It is forto depart the moder nombre into as many partis as ben of vnytees in the lasse nombre. And note wele that in makynge of dyvysiown ther ben .3. nombres necessary : that is to sey, the nombre to be dyvydede ; the nombre dyvydyng and the nombre exeant, other how oft, or quocient. Ay shalle the nombre that is to be dyvydede be more, other at the lest evene with the nombre the dyvysere, yf the nombre shalle be made by hole nombres. Therfor yf thow wolt any nombre dyvyde, write the nombre to be dyvydede in þe ouerer bordure by his differences, the dyvisere in the lower ordure by his differences, so that the last of the dyviser be vnder the last of the nombre to be dyvyde, the next last vnder the next last, and so of the others, yf it may competently be done ; as here :— *Definition of division.* *Dividend, Divisor, Quotient.* *How to set down your Sum.*

The residue			2	7
The quotient				5
To be dyvydede	3	4	2	
The dyvyser		6	3	

An example.

[1] Blank in MS.

Chapter VII. Division.

Examples.

Residuum			8 ‖		‖	2	7 ‖		2	6
Quociens		2	1 ‖	2	2 ‖		5 ‖			9
Diuidend*us*	6	8	0 ‖	6	6 ‖	3	4	2 ‖	3	2
Diuiser	3	2		‖ 3			6	3 ‖		3	4

When the last of the divisor must not be set below the last of the dividend.

And ther ben .2. causes whan the last figure may not be sette vnder the last, other that the last of the lower nombre may not be with-draw of the last of the ouerer nombre for it is lasse than the lower, other how be it, that it myght be with-draw as for hym-self fro 4 the ouerer the remenaunt may not so oft of them aboue, other yf þe last of the lower be euen to the figure aboue his hede, and þe next last oþer the figure be-fore þat be more þan the figure aboue sette. ¹These so ordeynede, me most wirche from the last figure of 8 þe nombre of the dyvyser, and se how oft it may be with-draw of and fro the figure aboue his hede, namly so that the remenaunt may be take of so oft, and to se the residue as here :—

¹ Fol. 53².

How to begin.

An example.

The residue		2	6
The quocient			9
To be dyvydede	3	3	2
The dyvyser		3	4

And note wele that me may not withe- 12 draw more than .9. tymes nether lasse than ones. Therfor se how oft þe figures of the lower ordre may be with-draw fro the figures of the ouerer, and the nombre that shewith þe 16 quocient most be writ ouer the hede of þat figure, vnder the whiche the first figure is, of the dyviser; And by that figure me most withe-draw alle oþer figures of the lower ordir and that of the figures aboue thaire hedis. This so done, me most sette forwarde þe figures 20 of the diuiser by o difference towardes the right honde and worche as before; and thus :—

Where to set the quotiente

Examples.

Residuum										.	1	2
quociens			6	5	4	‖			2	0	0	4
Diuidend*us*	3	5	5	1	2	2 ‖	8	8	6	3	7 ‖ 0	4
Diuisor		5	4	3		‖ 4	4	2	3			

The quocient			6	5	4
To be dyvydede	3	5	5	1	2	2
The dyvyser		5	4	3	

A special case.

And yf it happe after þe setlyng forwarde of the figures þat þe last of the divisor may not so oft be with-draw of the figure aboue 24 his hede, aboue þat figure vnder the whiche the first of the diuiser is writ me most sette a cifre in ordre of the nombre quocient, and sette the figures forwarde as be-fore be o difference alone, and so me shalle do in alle nombres to be dyvidede, for where the dyviser may 28

Chapter VIII. Progression.

not be wit*h*-draw me most sette there a cifre, and sette forwarde the figures; as here :—

							1	2
The residue							1	2
The quocient					2	0	0	4
To be dyvydede	8	8	6	3	7	0	4	
The dyvyser	4	4	2	3				

And me shalle not cesse fro suche settyng of figures forwarde, nether of settynge of þe quocient into the dyviser, neþer of subt*r*accioun of the dyvyser, tille the first of the dyvyser be wit*h*-draw fro þe first to be dividede. The whiche done, or ought,[1] oþer nought shalle remayne: and yf it be ought,[1] kepe it in the tables, And euer vny it to þe diviser. And yf þou wilt wete how many vnytees of þe divisio*u*n [2] wol growe to the nombre of the divisere, the nombre quocient wol shewe it: and whan suche divisio*u*n is made, and þou lust prove yf thow have wele done or no, Multiplie the quocient by the diviser, And the same figures wolle come ayene that thow haddest bifore and none other. And yf ought be residue, than wit*h* addicioun therof shalle come the same figures: And so multiplicacioun provithe divisioun, and dyvisio*u*n multiplicacio*u*n: as thus, yf multiplicacio*u*n be made, divide it by the multipliant, and the nombre quocient wol shewe the nombre that was to be multipliede, etc.

Progressio*u*n is of nombre after egalle excesse fro oone or tweyne take agregacio*u*n. of p*r*ogressio*u*n one is naturelle or con- tynuelle, þat oþer broken and discontynuelle. Naturelle it is, whan me begynnethe wit*h* one, and kepethe ordure oue*r*lepyng one; as .1. 2. 3. 4. 5. 6., etc., so þat the nombre folowynge passithe the other be-fore in one. Broken it is, whan me lepithe fro o nombre tille another, and kepithe not the contynuel ordire; as 1. 3. 5. 7. 9, etc. Ay me may begynne wit*h* .2., as þus; .2. 4. 6. 8., etc., and the nombre folowyng passethe the others by-fore by .2. And note wele, that naturelle p*r*ogressio*u*n ay begynnethe wit*h* one, and Inte*r*cise or broken p*r*ogressioun, omwhile begynnythe wit*h* one, omwhile wit*h* twayne. Of progressio*u*n naturell .2. rules ther be yove, of the whiche the first is this; whan the p*r*ogressioun naturelle endithe in even nombre, by the half therof multiplie þe next totalle oue*r*ere nombre; Example of grace: .1. 2. 3. 4. Multiplie .5. by .2. and so .10. comethe of, that is the totalle nombre þerof. The seconde rule is suche, whan the p*r*ogressioun naturelle endithe in nombre ode. Take the more porcio*u*n of the oddes, and multiplie therby the totalle nombre. Example of grace 1. 2. 3. 4. 5., multiplie

[1] 'nought' in MS.

.5. by .3, and thryes .5. shalle be resultant. so the nombre totalle
is .15. Of progresioun intercise, ther ben also .2.[1] rules; and þe
first is þis: Whan the Intercise progression endithe in even nombre
by half therof multiplie the next nombre to þat· halfe as .2.[1] 4. 6.   4
Multiplie .4. by .3. so þat is thryes .4., and .12. the nombre of alle
the progressioun, wolle folow. The seconde rule is this: whan the
progressioun interscise endithe in ode, take þe more porcioun of alle
þe nombre, [2]and multiplie by hym-selfe; as .1. 3. 5. Multiplie .3.   8
by hym-selfe, and þe some of alle wolle be .9., etc.

Here folowithe the extraccioun of rotis, and first in nombre
quadrates. Wherfor me shalle se what is a nombre quadrat,
and what is the rote of a nombre quadrat, and what it   12
is to draw out the rote of a nombre. And before other note
this divisioun: Of nombres one is lyneal, anoþer superficialle,
anoþer quadrat, anoþer cubike or hoole. lyneal is that þat is con-
sidrede after the processe, havynge no respect to the direccioun   16
of nombre in nombre, As a lyne hathe but one dymensioun that
is to sey after the lengthe. Nombre superficial is þat comethe
of ledynge of oo nombre into a-nother, wherfor it is callede superficial, for it hathe .2. nombres notyng or mesurynge hym, as a   20
superficialle thynge hathe .2. dimensions, þat is to sey lengthe and
brede. And for bycause a nombre may be hade in a-nother by .2.
maners, þat is to sey other in hym-selfe, oþer in anoþer, Vnderstonde yf it be had in hym-self, It is a quadrat. ffor dyvisioun   24
write by vnytes, hathe .4. sides even as a quadrangille. and yf the
nombre be hade in a-noþer, the nombre is superficiel and not
quadrat, as .2. hade in .3. makethe .6. that is þe first nombre super-
ficielle; wherfor it is open þat alle nombre quadrat is superficiel,   28
and not conuertide. The rote of a nombre quadrat is þat nombre
that is had of hym-self, as twies .2. makithe 4. and .4. is the first
nombre quadrat, and 2. is his rote. 9. 8. 7. 6. 5. 4. 3. 2. 1. / The
rote of the more quadrat .3. 1. 4. 2. 6. The most nombre quadrat   32
9. 8. 7. 5. 9. 3. 4. 7. 6. / the remenent ouer the quadrat .6. 0. 8.
4. 5. / The first caas of nombre quadrat .5. 4. 7. 5. 6. The rote .2.
3. 4. The seconde caas .3. 8. 4. 5. The rote .6. 2. The thirde
caas .2. 8. 1. 9. The rote .5. 3. The .4. caas .3. 2. 1. The rote   36
.1. 7. / The 5. caas .9. 1. 2. 0. 4. / The rote 3. 0. 2. The solide
nombre or cubike is þat þat comythe of double ledyng of nombre
in nombre; And it is clepede a solide body that hathe þer-in .3

---

[1] 3 written for 2 in MS.

Chapter X. Extraction of Square Root.

[dimensions] þat is to sey, lengthe, brede, and thiknesse. so þat nombre hathe .3. nombres to be brought forthe in hym. Buȝ nombre may be hade twies in nombre, for other it is hade in hym-selfe, oþer in a-noþer. If a nombre be hade twies in hym-self, oþer ones in his quadrat, þat is the same, þat a cubike ¹is, And is the same that is solide. And yf a nombre twies be hade in a-noþer, the nombre is clepede solide and not cubike, as twies .3. and þat .2. makithe .12. Wherfor it is opyne that alle cubike nombre is solide, and not conuertide. Cubike is þat nombre þat comythe of ledynge of hym-selfe twyes, or ones in his quadrat. And here-by it is open that o nombre is the roote of a quadrat and of a cubike. Natheles the same nombre is not quadrat and cubike. Opyne it is also that alle nombres may be a rote to a quadrat and cubike, but not alle nombre quadrat or cubike. Therfor sithen þe ledynge of vnyte in hym-self ones or twies nought comethe but vnytes, Seithe Boice in Arsemetrike, that vnyte potencially is al nombre, and none in act. And vndirstonde wele also that betwix euery .2. quadrates ther is a

Three dimensions of solids.

¹ *Fol. 54. Cubic numbers.*

All cubics are solid numbers.

No number may be both linear and solid.

Unity is not a number.

Residuum			0	‖			4	‖		0				0	
Quadrande	4	3	5	6 ‖ 3	0	2	9 ‖ 1	7	4	2	4 ‖ 1	9	3	6
Duplum	1	2	~		‖ 1	0		‖ 2		6		‖	[8]	²	
Subduplum		6		6 ‖		5		5 ‖ 1		3		2		4		4

Examples of square roots.

meene proporcionalle, That is openede thus; lede the rote of o quadrat into the rote of the oþer quadrat, and þan wolle þe meene shew. Also betwix the next .2. cubikis, me may fynde a double meene, that is to sey a more meene and a lesse. The more meene thus, as to brynge the rote of the lesse into a quadrat of the more. The lesse thus, If the rote of the more be brought Into the quadrat of the lesse.

A note on mean proportionals.

³To draw a rote of the nombre quadrat it is What-euer nombre be proposede to fynde his rote and to se yf it be quadrat. And yf it be not quadrat the rote of the most quadrat fynde out, vnder the nombre proposede. Therfor yf thow wilt the rote of any quadrat nombre draw out, write the nombre by his differences, and compt the nombre of the figures, and wete yf it be ode or even. And yf it be even, than most thow begynne worche vnder the last save one. And yf it be ode with the last; and forto sey it shortly, al-weyes fro the last ode me shalle begynne. Therfor vnder the last in an od place sette, me most fynde a digit, the whiche lade in hym-selfe it puttithe away that, þat is ouer his hede, oþer as neighe as me

To find a square root.

Begin with the last odd place.

² 7 in MS.   ³ runs on in MS.

Chapter X. Extraction of Square Root.

Find the nearest square root of that number, subtract,
may: suche a digit founde and with-draw fro his ouere, me most double that digit and sette the double vnder the next figure towarde the right honde, and his vnder double vnder hym. That done, than

double it,
me most fynde a-noþer digit vnder the next figure bifore the doublede, 4

[1] Fol. 54 b.
and set the double one to the right.
Find the second figure by division.
the whiche [1] brought in double settethe a-way alle that is ouer his hede as to rewarde of the doublede: Than brought into hym-self settithe all away in respect of hym-self, Other do it as nye as it may be do: other me may with-draw the digit [2][last] founde, and 8

Multiply the double by the second figure, and add after it the square of the second figure, and subtract.
lede hym in double or double hym, and after in hym-selfe; Than Ioyne to-geder the produccione of them bothe, So that the first figure of the last product be addede before the first of the first productes, the seconde of the first, etc. and so forthe, subtrahe fro the totalle 12 nombre in respect of þe digit. And if it hap þat no digit may be

Examples.

The residue					‖			‖		5	4	3	2
To be quadrede	4	1	2	0	9 ‖	1	5	1 ‖	3	9, ‖ 9	0	0	5	4	3	2
The double		4	0		‖	2		4 ‖		6		0		0
The vnder double	2		0		3 ‖ 1		2		3 ‖[3]‖	‖[0]‖	‖[0]‖	0

founde, Than sette a cifre vndre a cifre, and cesse not tille thow fynde a digit; and whan thow hast founde it to double it, neþer to

Special cases.
sette the doublede forwarde nether the vnder doublede, Till thow 16 fynde vndre the first figure a digit, the whiche lade in alle double, settyng away alle that is ouer hym in respect of the doublede: Than lede hym into hym-selfe, and put a-way alle in regarde of hym, other

The residue.
as nyghe as thow maist. That done, other ought or nought wolle 20 be the residue. If nought, than it shewithe that a nombre componede was the quadrat, and his rote a digit last founde with vndere-double other vndirdoubles, so that it be sette be-fore: And yf ought [3] remayne, that shewith that the nombre proposede was not 24 quadrat,[4] but a digit [last found with the subduple or subduples

This table is constructed for use in cube root sums, giving the value of ab.[5]

1	2	3	4	5	6	7	8	9
2	8	12	16	20	24	28	32	36
3	18	27	36	45	54	63	72	81
4	32	48	64	80	96	112[5]	128	144
5	50	75	100	125	150	175	200	225
6	72	108	144	180	216	252	288	324
7	98	147	196	245	294	343	393	441
8	128	192	256	320	384	448	512	576
9	168	243	324	405	486	567	648	729[6]

[2] 'so' in MS.   [3] 'nought' in MS.
[4] MS. adds here: 'wher-vpone se the table in the next side of the next leefe.'
[5] 110 in MS.   [6] 0 in MS.

Chapter XI. Extraction of Cube Root.

is] The rote of the most quadrat conteynede vndre the nombre proposede. Therfor yf thow wilt prove yf thow have wele do or no, Multiplie the digit last founde with the vnder-double oþer vnder-doublis, and thow shalt fynde the same figures that thow haddest before; And so that nought be the ¹residue. And yf thow have any residue, than with the addicioun þerof that is reserued with-out in thy table, thow shalt fynde thi first figures as thow haddest them before, *etc.*

How to prove the square root without or with a remainder.
¹ Fol. 55.

Heere folowithe the extraccioun of rotis in cubike nombres; wher-for me most se what is a nombre cubike, and what is his roote, And what is the extraccioun of a rote. A nombre cubike it is, as it is before declarede, that comethe of ledyng of any nombre twies in hym-selfe, other ones in his quadrat. The rote of a nombre cubike is the nombre that is twies lade in hym-selfe, or ones in his quadrat. Wher-thurghe it is open, that euery nombre quadrat or cubike have the same rote, as it is seide before. And forto draw out the rote of a cubike, It is first to fynde þe nombre proposede yf it be a cubike; And yf it be not, than thow most make extraccioun of his rote of the most cubike vndre the nombre proposide his rote founde. Therfor proposede some nombre, whos cubical rote þou woldest draw out; First thow most compt the figures by fourthes, that is to sey in the place of thousandes; And vnder the last thousande place, thow most fynde a digit, the whiche lade in hym-self cubikly puttithe a-way that þat is ouer his hede as in respect of hym, other as nyghe as thow maist. That done, thow most treblille the digit, and that triplat is to be put vnder the .3. next figure towarde the right honde, And the vnder-trebille vnder the trebille; Than me most fynde a digit vndre the next figure bifore the triplat, the whiche with his vnder-trebille had into a trebille, afterwarde other vnder[trebille]² had in his produccioun, putteþe a-way alle that is ouer it in regarde of³ [the triplat. Then lade in hymself puttithe away that þat is over his hede as in respect of hym, other as nyghe as thou maist :] That done, thow most treblille the digit ayene, and the triplat is to be sette vnder the next .3. figure as before, And the vnder-trebille vnder the trebille: and than most thow sette forwarde the first triplat with his vndre-trebille by .2. differences. And than most thow fynde a digit vnder the next figure before the triplat, the whiche withe his vnder-triplat had in his triplat after-

Definition of a cubic number and a cube root.

Mark off the places in threes.
Find the first digit;
treble it and place it under the next but one, and multiply by the digit.
Then find the second digit.
Multiply the first triplate and the second digit, twice by this digit.

² double in MS.   ³ 'it hym-selfe' in MS.

NOMBRYNGE.   E

Chapter XI. Extraction of Cube Root.

Subtract.
1 Fol. 55 b.

warde, other vnder-treblis lad in product [1]It sittethe a-way all that is ouer his hede in respect of the triplat than had in hym-self cubikly,[2] or as nyghe as ye may.

Examples.

Residuum						5 ‖					4 ‖	1	0	1	9
Cubicandus	8	3	6	5	4	3	2 ‖ 3	0	0	7	6	7 ‖ 1	1	6	6	7
Triplum			6	0		‖			1	8	‖			4	
Subtriplum	2			0		[3] ‖		6			7 ‖	2			2

Continue this process till the first figure is reached.

Nother me shalle not cesse of the fyndynge of that digit, neither of 4 his triplacioun, ne}er of the triplat-is [3]anterioracioun, that is to sey, settyng forwarde by .2. differences, Ne therof the vndre-triple to be put vndre the triple, Nether of the multiplicacioun þerof, Neither of the subtraccioun, tille it come to the first figure, vnder the 8 whiche is a digitalle nombre to be founde, the whiche withe his vndre-treblis most be hade in tribles, After-warde without vnder-treblis to be hade into produccioun, settyng away alle that is ouer the hede of the triplat nombre, After had into hymselfe cubikly, 12 and sette alle-way that is ouer hym. Also note wele that the produccion com-

Examples.

To be cubicede	1	7	2	8 ‖ 3	2	7	6	8
The triple			3	2 ‖			9	
The vnder triple			1	2 ‖	[3]‖		3	3

ynge of the ledyng of a digite founde[4] me may adde to, and also with-draw fro of the totalle nombre sette above that digit so **The residue.** founde.[5] That done ought or nought most be the residue. If it be nought, It is open that the nombre proposede was a cubike 16 nombre, And his rote a digit founde last with the vnder-triples: If the rote therof wex bade in hym-selfe, and afterwarde product they shalle make the first figures. And yf ought be in residue, kepe that without in the table; and it is opene that the nombre was not 20 a cubike. but a digit last founde with the vndirtriplis is rote of the most cubike vndre the nombre proposede conteynede, the **Special cases.** whiche rote yf it be hade in hym-selfe, And afterwarde in a product of that shalle growe the most cubike vndre the nombre proposede 24 conteynede, And yf that be addede to a cubike the residue reseruede *6 Fol. 56.* in the table, wolle make the same figures that ye hade first. [6]And

---

[2] MS. adds here: 'it settethe a-way alle his respect.'
[3] 'aucterioracioun' in MS.
[4] MS. adds here: 'with an vndre-triple / other of an vndre-triple in a triple or triplat is And after-warde with out vndre-triple other vndre-triplis in the product and ayene that product that comethe of the ledynge of a digit founde in hym-selfe cubicalle' /
[5] MS. adds here: 'as ther had be a divisioun made as it is openede before.'

Table of Numbers, &c.

yf no digit after the anterioracioun[1] may not be founde, than put there a cifre vndre a cifre yndir the thirde figure, And put forwarde þe figures. Note also wele that yf in the nombre proposede ther ben no place of thowsandes, me most begynne vnder the first figure in the extraccioun of the rote. some vsen forto distingue the nombre by threes, and ay begynne forto wirche vndre the first of

Special case.

The residue							0 ‖					1	1
The cubicandus	8	0	0	0	0	0	0 ‖ 8	2	4	2	4	1	9
The triple			²	0	0		‖		6				
The vndertriple	[2]			0	0		‖ 2		6	2			

Examples.

the last ternary other uncomplete nombre, the whiche maner of operacioun accordethe with that before. And this at this tyme suffisethe in extraccioun of nombres quadrat or cubikes etc.

    1   2   3   4   5   6
one. x. an. hundrede / a thowsande / x. thowsande / An hundrede
                  7
thowsande / A thowsande tymes a thowsande / x. thousande tymes a thousande / An hundrede thousande tymes a thousande A thousande thousande tymes a thousande / this is the x place etc.

A table of numbers; probably from the Abacus.

[Ende.]

[1] MS. anteriocacioun.   [2] 4 in MS.

Accomptynge by counters.

[1 116 b.] [1] ¶ The seconde dialoge of accomptynge by counters.

Mayster.

NOwe that you haue learned the commen kyndes of Arithmetyke with the penne, you shall se the same art in counters: whiche feate doth not only serue for them that can not write and rede, but also for them that can do bothe, but haue not at some tymes theyr penne or tables redye with them. This sorte is in two fourmes commenly. The one by lynes, and the other without lynes: in that y^t hath lynes, the lynes do stande for the order of places: and in y^t that hath no lynes, there must be sette in theyr stede so many counters as shall nede, for eche lyne one, and they shall supplye the stede of the lynes. *S.* By examples I shuld better perceaue your meanynge. *M.* For example of the ly^{2}nes: Lo here you se .vi. lynes whiche stande for syxe places so that the nethermost standeth for ye fyrst place, and the next aboue it, for the second: and so vpward tyll you come to the hyghest, which is the syxte lyne, and standeth for the syxte place. Now what is the valewe of euery place or lyne, you may perceaue by the figures whiche I haue set on them, which is accordynge as you learned before in the Numeration of figures by the penne: for the fyrste place is the place of vnities or ones, and euery counter set in that lyne betokeneth but one: *and* the seconde lyne is the place of 10, for euery counter there, standeth for 10. The thyrd lyne the place of hundredes: the fourth of thousandes: *and* so forth. *S.* Syr I do perceaue that the same order is here of lynes, as was in the other figures [3] by places, so that you shall not nede longer to stande about Numeration, excepte there be any other difference. *M.* Yf you do vnderstande it, then how wyll you set 1543? *S.* Thus, as I suppose. *M.* You haue set ye places truely, but your figures be not mete for this vse:

[2] 117 a

Numeration.

[3] 117 b.

Addition on the Counting Board. 53

for the metest figure in this behalfe, is the figure of a counter round, as you se here, where I haue expressed that same summe. *S.* So that you haue not one figure for 2, 4 nor 3, nor 4, and so forth, but as many digettes as you haue, you set in the lowest lyne: and for euery 10 you set one in the second line: and so of other. But I know not by what reason you set that one counter for 500 betwene two lynes. *M.* you shall remember this, that when so euer you nede to set downe 5, 50, or 500, or 5000, or so forth any other nomber, whose numerator [1]is 5, you shall set one counter for it, in the next space aboue the lyne that it hath his denomination of, as in this example of that 500, bycause the numerator is 5, it must be set in a voyd space: and bycause the denominator is hundred, I knowe that his place is the voyde space next aboue hundredes, that is to say, aboue the thyrd lyne. And farther you shall marke, that in all workynge by this sorte, yf you shall sette downe any summe betwene 4 and 10, for the fyrste parte of that nomber you shall set downe 5, & then so many counters more, as there reste nombers aboue 5. And this is true bothe of digettes and articles. And for example I wyll set downe this summe 287965, which summe yf you marke well, you nede none other examples for to lerne the numeration of [2]this forme. But this shal you marke, that as you dyd in the other kynde of arithmetike, set a pricke in the places of thousandes, in this worke you shall sette a starre, as you se here. *S.* Then I perceaue numeration, but I praye you, howe shall I do in this arte to adde two summes or more together? *M.* The easyest way in this arte is, to adde but 2 summes at ones together: how be it you may adde more, as I wyll tell you anone. Therfore when you wyll adde two summes, you shall fyrst set downe one of them, it forseth not whiche, *and* then by it drawe a lyne crosse the other lynes. And afterward set downe the other summe, so that that lyne may be betwene them, as yf you wolde adde 2659 to 8342, you must set your summes as you se here. And then yf you lyst, you [3]may adde the one to the other in the same place, or els you may adde them both together in a newe place: which waye, bycause it is moste playnest, I wyll showe you fyrst. Therfore wyl I begynne at the vnites, whiche in the fyrst summe is but 2, *and* in ye second summe 9, that maketh 11, those do I take vp, and for them I set 11 in the new roume, thus,

[1] 118 a.

[2] 118 b.

Addition.

[3] 119 a.

Then do I take vp all ye articles vnder a hundred, which in the fyrst summe are 40, and in the second summe 50, that maketh 90: or you may saye better, that in the fyrste summe there are 4 articles of 10, and 4 in the seconde summe 5, which make 9, but then take hede that you sette them in theyr [1]ryght lynes as you se here. Where I haue taken awaye 40 from the fyrste summe, and 50 from ye 8 second, and in theyr stede I haue set 90 in the thyrde, whiche I haue set playnely y^t you myght well perceaue it: how be it seynge that 90 with the 10 that was in ye thyrd roume all redy, doth make 100, I myghte better for those 6 counters set 1 in the thyrde 12 lyne, thus: For it is all one summe as you may se, but it is beste, neuer to set 5 counters in any line, for that may be done with 1 counter in a hygher place. *S.* I iudge that good reason, for many are vnnedefull, where one wyll serue. 16

M. Well, then [2]wyll I adde forth of hundredes: I fynde 3 in the fyrste summe, and 6 in the seconde, whiche make 900, them do I take vp and set in the thyrd roume where is one hundred all redy, to whiche I put 900, and it wyll be 1000, therfore I set one 20 counter in the fourth lyne for them all, as you se here. Then adde I ye thousandes together, whiche in the fyrst summe are 8000, *and* in ye second 2000, that maketh 10000: them do I take vp from those 24 two places, and for them I set one counter in the fyfte lyne, and then appereth as you se, to be 11001, for so many doth amount of the addition of 8342 to 2659. [3]*S.* Syr, this I do perceaue: but how shall I set one summe to an other, not 28 chaungynge them to a thyrde place? *M.* Marke well how I do it: I wyll adde together 65436, and 3245, whiche fyrste I set downe thus. Then do I begynne with the smalest, which 32 in the fyrst summe is , that do I take vp, and wold put to the other 5 in the seconde summe, sauynge that two counters can not be set in a voyd place of 5, but for them bothe I must set 1 in the seconde lyne, which is the place of 10, therfore I take vp the 5 of 36 the fyrst summe, *and* the 5 of the seconde, and for them I set 1 in the second lyne, [4]as you se here. Then do I lyke wayes take vp the 4 counters of the fyrste summe *and* 40

Subtraction on the Counting Board.

seconde lyne (which make 40) and adde them to the 4 counters of the same lyne, in the second su*m*me, and it maketh 80, But as I sayde I maye not conueniently set aboue 4 counters in one lyne, 4 therfore to those 4 that I toke vp in the fyrst su*m*me, I take one also of the seconde su*m*me, and then haue I taken vp 50, for whiche 5 counters I sette downe one in the space ouer ye second lyne, as here doth appere. ———— [1]and then is there 80, 8 as well w^t those 4 counters, as yf I had set downe ye other 4 also. Now do I take the 200 in the fyrste su*m*me, and adde them to the 400 in the seconde summe, and it maketh 600, therfore I take vp the 2 12 counters in the fyrste summe, and 3 of them in the seconde summe, and for them 5 I set 1 in ye space aboue, thus. Then I take ye 3000 in ye fyrste su*m*me, vnto whiche there are none in the 16 second summe agreynge, therfore I do onely remoue those 3 counters from the fyrste summe into the seconde, as here doth appere. [2]And so you see the hole su*m*me, that amou*n*teth of the addytio*n* of 65436 with 3245 to be 6868[1]. 20 And yf you haue marked these two exa*m*ples well, you nede no farther enst*r*uctio*n* in Addition of 2 only su*m*mes: but yf you haue more then two summes to adde, you may adde them thus. Fyrst adde two of them, and then adde the thyrde, 24 and ye fourth, or more yf there be so many: as yf I wolde adde 2679 with 4286 and 1391. Fyrste I adde the two fyrste summes thus. [3]And then I adde the thyrde thereto thus. 28 And so of more yf you haue them. *S.* Nowe I thynke beste that you passe forth to Subtraction, except there be any wayes to examyn this maner of Addition, then I thynke that were 32 good to be knowen nexte. *M.* There is the same profe here that is in the other Add*i*tion by the penne, I meane Subtraction, for that onely is a sure waye: but consyderynge that Subtraction must be fyrste knowen, I wyl fyrste teache you the arte of Subtraction, and 36 that by this example: I wolde subtracte 2892 out of 8746. These summes must I set downe as I dyd in Addition: but here it is best [4]to set the lesser no*m*ber fyrste, thus. Then shall I begynne to sub- 40 tracte the greatest nombres fyrste (contrary to the vse of the penne) y^t is the thousandes in this example: therfore I fynd amongest the thousandes 2, for which I withdrawe so many from the seconde summe (where are 8) and so remayneth there 6, as this example showeth. Then do I lyke wayes with the hundredes, of whiche in the fyrste summe [1]I fynde 8, and is the seconde summe but 7, out of whiche I can not take 8, therfore thus muste I do: I muste loke how moche my summe dyffereth from 10, whiche I fynde here to be 2, then must I bate for my summe of 800, one thousande, and set downe the excesse of hundredes, that is to saye 2, for so moche 100[0} is more then I shuld take vp. Therfore from the fyrste summe I take that 800, and from the second summe where are 6000, I take vp one thousande, and leue 5000; but then set I downe the 200 unto the 700 y^t are there all redye, and make them 900 thus. Then come I to the articles of tennes where in the fyrste summe I fynde 90, [2]and in the seconde summe but only 40: Now consyderyng that 90 can not be bated from 40, I loke how moche y^t 90 doth dyffer from the next summe aboue it, that is 100 (or elles whiche is all to one effecte, I loke how moch 9 doth dyffer from 10) and I fynd it to be 1, then in the stede of that 90, I do take from the second summe 100: but consyderynge that it is 10 to moche, I set downe 1 in ye nexte lyne beneth for it, as you se here. Sauynge that here I haue set one counter in ye space in stede of 5 in ye nexte lyne. And thus haue I subtracted all saue two, which I must bate from the 6 in the second summe, and there wyll remayne 4, thus. So y^t yf I subtracte 2892 from 8746, the remayner wyll be 5854, [3]And that this is truely wrought, you maye proue by Addition: for yf you adde to this remayner the same summe that you dyd subtracte, then wyll the formar summe 8746 amount agayne. S. That wyll I proue: and fyrst I set the summe that was subtracted, which was 2892, and then the remayner 5854, thus. Then do I adde fyrst ye 2 to 4, whiche maketh 6, so take I vp 5 of those counters, and in theyr stede I sette 1 in the space, as here appereth. [4]Then do I adde the 90 nexte aboue to the 50, and it maketh 140, therfore I take vp those 6 counters, and for them I sette 1 to the hundredes in ye thyrde lyne, and 4 in y^e

Subtraction by Counters.

second lyne, thus. Then do I come to the hundredes, of whiche I fynde 8 in the fyrst summe, and 9 in ye second, that maketh 1700, therfore I take vp those 9 counters, and in theyr stede I sette 1 in the .iiii. lyne, and 1 in the space nexte beneth, and 2 in the thyrde lyne, as you se here. Then is there lefte in the fyrste summe but only 2000, whiche I shall take vp from thence, and set [1]in the same lyne in ye second summe, to ye one y^t is there all redy: *and* then wyll the hole summe appere (as you may wel se) to be 8746, which was ye fyrst grosse summe, *and* therfore I do perceaue, that I hadde well subtracted before. And thus you may se how Subtraction maye be tryed by Addition. *S.* I perceaue the same order here w^t counters, y^t I lerned before in figures. *M.* Then let me se howe can you trye Addition by Subtraction. *S.* Fyrste I wyl set forth this example of Addition where I haue added 2189 to 4988, and the hole summe appereth to be 7177, [2]Nowe to trye whether that summe be well added or no, I wyll subtract one of the fyrst two summes from the thyrd, and yf I haue well done ye remayner wyll be lyke that other summe. As for example: I wyll subtracte the fyrste summe from the thyrde, whiche I set thus in theyr order. Then do I subtract 2000 of the fyrste summe from ye second summe, and then remayneth there 5000 thus. Then in the thyrd lyne, I subtract ye 100 of the fyrste summe, from the second summe, where is onely 100 also, and then in ye thyrde lyne resteth nothyng. Then in the second lyne with his space ouer hym, I fynde 80, which I shuld subtract [3]from the other summe, then seyng there are but only 70 I must take it out of some hygher summe, which is here only 5000, therfore I take vp 5000, and seyng that it is to moch by 4920, I sette·downe so many in the seconde roume, whiche with the 70 beynge there all redy do make 4990, & then the summes doth stande thus. Yet remayneth there in the fyrst summe 9, to be bated from the second summe, where in that place of vnities dothe appere only 7, then I muste bate a hygher summe, that is to saye 10, but seynge that 10 is more then 9 (which I shulde abate) by 1, therfore shall I take vp one counter from the seconde lyne, *and* set downe the same in the fyrst [4]or

[1] 118 b.

[2] 119 a.

[3] 119 b.

[4] 120 a.

Multiplication by Counters.

lowest lyne, as you se here. And so haue I ended this worke, *and* the summe appereth to be ye same, whiche was ye seconde summe of my addition, and therfore I perceaue, I haue wel done. *M.* To stande longer about this, it is but folye: excepte that this you maye also vnderstande, that many do begynne to subtracte with counters, not at the hyghest summe, as I haue taught you, but at the nethermoste, as they do vse to adde: and when the summe to be abatyd, in any lyne appeareth greater then the other, then do they borowe one of the next hygher roume, as for example: yf they shuld abate 1846 from 2378, they set ye summes thus. [1]And fyrste they take 6 whiche is in the lower lyne, and his space from 8 in the same roumes, in ye second summe, and yet there remayneth 2 counters in the lowest lyne. Then in the second lyne must 4 be subtracte from 7, and so remayneth there 3. Then 8 in the thyrde lyne and his space, from 3 of the second summe can not be, therfore do they bate it from a hygher roume, that is, from 1000, and bycause that 1000 is to moch by 200, therfore must I sette downe 200 in the thyrde lyne, after I haue taken vp 1000 from the fourth lyne: then is there yet 1000 in the fourth lyne of the fyrst summe, whiche yf I withdrawe from the seconde summe, then doth all ye figures stande in this order.

So that (as you se) it differeth not greatly whether you begynne subtraction at the hygher lynes, or at [2]the lower. How be it, as some menne lyke the one waye beste, so some lyke the other: therfore you now knowyng bothe, may vse whiche you lyst. But nowe touchynge Multiplication: you shall set your nombers in two roumes, as you dyd in those two other kyndes, but so that the multiplier be set in the fyrste roume. Then shall you begyn with the hyghest nombers of ye seconde roume, and multiply them fyrst after this sort. Take that ouermost lyne in your fyrst workynge, as yf it were the lowest lyne, setting on it some mouable marke, as you lyste, and loke how many counters be in hym, take them vp, and for them set downe the hole multyplyer, so many tymes as you toke vp counters, reckenyng, I saye that lyne for the vnites: *and* when you haue so done with the hygheest nomber then come to the nexte lyne beneth, *and* do euen so with it, and so with ye next, tyll you haue done all. And yf there be any nomber in a space, then for it [3]shall you take ye multiplyer 5 tymes, and then must you recken that lyne for the vnites whiche is nexte beneth that space: or els

Multiplication by Counters.

after a shorter way, you shall take only halfe the multyplyer, but then shall you take the lyne nexte aboue that space, for the lyne of vnites: but in suche workynge, yf chaunce your multyplyer be an odde nomber, so that you can not take the halfe of it iustly, then muste you take the greater halfe, and set downe that, as if that it were the iuste halfe, and farther you shall set one counter in the space beneth that line, which you recken for the lyne of vnities, or els only remoue forward the same that is to be multyplyed. *S.* Yf you set forth an example hereto I thynke I shal perceaue you. *M.* Take this example: I wold multiply 1542 by 365, therfore I set ye nombers thus. [diagram] ¹Then fyrste I begynne at the 1000 in [diagram] ye hyghest roume, as yf it were ye fyrst place, & I take it vp, settynge downe for it so often (that is ones) the multyplyer, which is 365, thus, as you se here: [diagram] where for the one counter taken vp from the fourth lyne, I [diagram] haue sette downe other 6, whiche make ye summe of the multyplyer, reckenynge that fourth lyne, as yf it were the fyrste: whiche thyng I haue marked by the hand set at the begynnyng of ye same, *S.* I perceaue this well: for in dede, this summe that you haue set downe is 365000, for so moche doth amount ²of 1000, multiplyed by 365. *M.* Well then to go forth, in the nexte space I fynde one counter which I remoue forward but take not vp, but do (as in such case I must) set downe the greater halfe of my multiplier (seyng it is an odde nomber) which is 182, and here I do styll let that fourth place stand, as yf it were ye fyrst: as in this fourme [diagram] you se, where I haue set this multiplycation with ye other: but for the ease of your vnderstandynge, I haue set a lytell lyne betwene them: now shulde they both in one summe stand thus. ³Howe be it an other fourme to multyplye suche counters [diagram] in space is this: Fyrst to remoue the fynger to the lyne nexte benethe ye space, and then to take vp ye counter, and to set downe ye multiplyer .v. tymes, as here you se. Which summes yf you do

[diagram]

adde together into one summe, you shal perceaue that it wyll be ye

[1] 122 *a.*

[2] 122 *b.*

[3] 123 *a.*

Multiplication by Counters.

[1] 123 b. same y^t appeareth of ye other working before, so that [1]bothe sortes are to one entent, but as the other is much shorter, so this is playner to reason, for suche as haue had small exercyse in this arte. Not withstandynge you maye adde them in your mynde before you sette them downe, as in this example, you myghte haue sayde 5 tymes 300 is 1500, and 5 tymes 60 is 300, also 5 tymes 5 is 25, whiche all put together do make 1825, which you maye at one tyme set downe yf you lyste. But nowe to go forth, I must remoue the hand to the nexte counters, whiche are in the second lyne, and there must I take vp those 4 counters, settynge downe for them my multiplyer 4 tymes, whiche thynge other I maye do at 4 tymes seuerally, or elles I may gather that hole summe in my mynde fyrste, and then set it downe: as to saye 4 tymes 300 is 1200: 4 tymes 60 are 240: and 4 tymes 5 make 20: y^t is in all 1460, y^t shall I set downe also: as here you

[2] 124 a. se. [2]whiche yf I ioyne
in one summe with the formar nombers, it wyll appeare thus.

Then to ende this multiplycation, I remoue the fynger to the lowest lyne, where are onely 2, them do I take vp,
and in theyr stede do I set downe twyse 365, that is 730, for

[3] 124 b. which I set [3]one in the space aboue the thyrd lyne for 500, and 2 more in the thyrd lyne with that one that is there all redye, and the reste in theyr order, and so haue I ended the hole summe thus.

Wherby you se, that 1542 (which is the nomber of yeares syth Ch[r]ystes incarnation) beyng multyplyed by 365 which is the nomber of dayes in one yeare) dothe amounte vnto 562830, which declareth ye nomber of daies sith Chrystes incarnation vnto the ende of 1542[4] yeares. (besyde 385 dayes and 12 houres for lepe yeares). S. Now wyll I proue by an other example,

[5] 125 a. as this: 40 labourers (after 6d. ye day for eche man) haue wrought 28 dayes, I wold [5]know what theyr wages doth amount vnto: In this case muste I worke doublely: fyrst I must multyplye the nomber of the labourers by ye wages of a man for one day, so wyll ye charge of one daye amount: then secondarely shall I multyply that charge of one daye, by the hole nomber of dayes, and so wyll the hole summe appeare: fyrst therefore I shall set the summes thus.

[4] 1342 in original.

Division on the Counting Board.

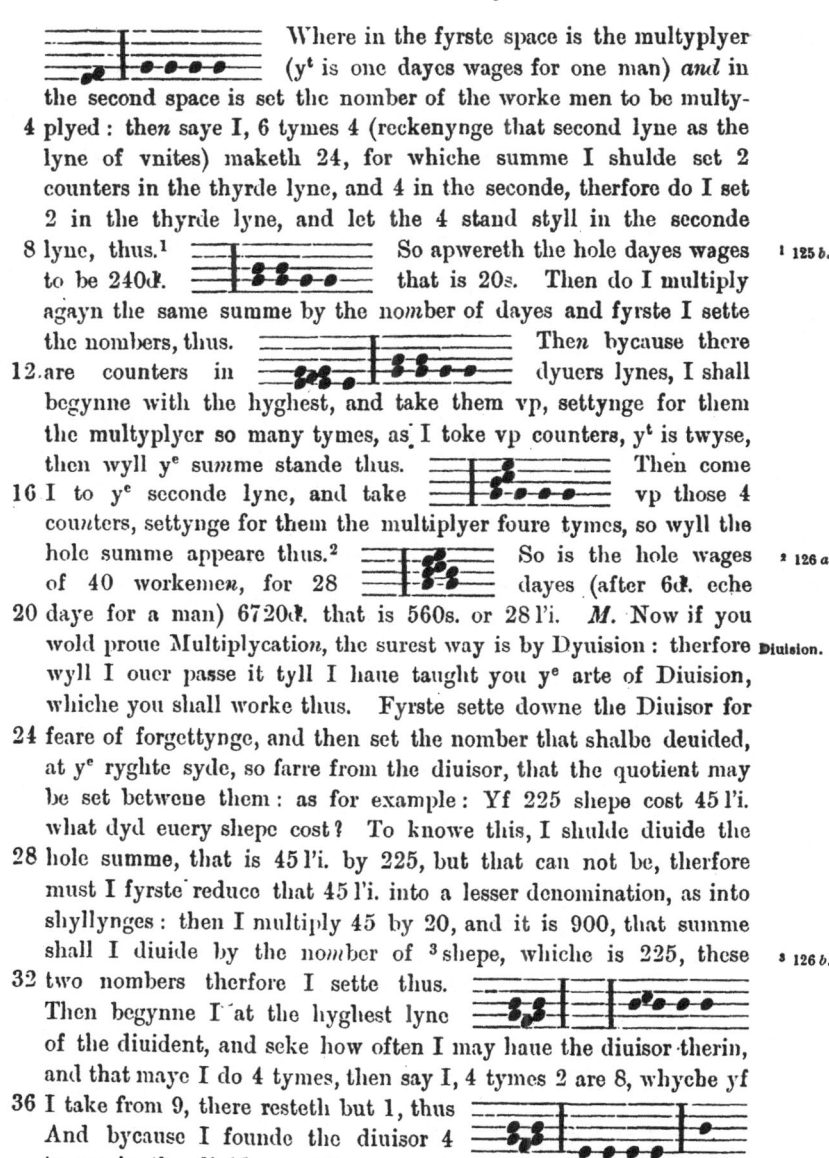

Where in the fyrste space is the multyplyer (y^t is one dayes wages for one man) *and* in the second space is set the nomber of the worke men to be multy-
4 plyed: the*n* saye I, 6 tymes 4 (reckenynge that second lyne as the lyne of vnites) maketh 24, for whiche summe I shulde set 2 counters in the thyrde lyne, and 4 in the seconde, therfore do I set 2 in the thyrde lyne, and let the 4 stand styll in the seconde
8 lyne, thus.[1]   So apwereth the hole dayes wages   [1] 125 b. to be 240d.   that is 20s. Then do I multiply agayn the same summe by the no*m*ber of dayes and fyrste I sette the nombers, thus.   Then bycause there
12 are counters in   dyuers lynes, I shall begynne with the hyghest, and take them vp, settynge for them the multyplyer so many tymes, as I toke vp counters, y^t is twyse, then wyll ye su*m*me stande thus.   Then come
16 I to ye seconde lyne, and take   vp those 4 cou*n*ters, settynge for them the multiplyer foure tymes, so wyll the hole summe appeare thus.[2]   So is the hole wages   [2] 126 a. of 40 workeme*n*, for 28   dayes (after 6d. eche
20 daye for a man) 6720d. that is 560s. or 28 l'i. *M.* Now if you wold proue Multiplycatio*n*, the surest way is by Dyuision: therfore   Diuision. wyll I ouer passe it tyll I haue taught you ye arte of Diuision, whiche you shall worke thus. Fyrste sette downe the Diuisor for
24 feare of forgettynge, and then set the nomber that shalbe deuided, at ye ryghte syde, so farre from the diuisor, that the quotient may be set betwene them: as for example: Yf 225 shepe cost 45 l'i. what dyd euery shepe cost? To knowe this, I shulde diuide the
28 hole summe, that is 45 l'i. by 225, but that can not be, therfore must I fyrste reduce that 45 l'i. into a lesser denomination, as into shyllynges: then I multiply 45 by 20, and it is 900, that summe shall I diuide by the no*m*ber of [3] shepe, whiche is 225, these   [3] 126 b.
32 two nombers therfore I sette thus.   Then begynne I at the hyghest lyne of the diuident, and seke how often I may haue the diuisor therin, and that maye I do 4 tymes, then say I, 4 tymes 2 are 8, whyche yf
36 I take from 9, there resteth but 1, thus   And bycause I founde the diuisor 4   tymes in the diuidente, I haue set (as you se) 4 in the myddle roume, which [4] is the place of the quotient: but now must I take   [4] 127 a.
40 the reste of the diuisor as often out of the remayner: therfore come

Division by Counters.

I to the seconde lyne of the diuisor, sayeng 2 foure tymes make 8, take 8 from 10, *and* there resteth 2, thus. Then come I to the lowest nomber, which is 5, and multyply it 4 tymes, so is it 20, that take I from 20, and there remayneth nothynge, so that I se my quotient to be 4, whiche are in valewe shyllynges, for so was the diuident: and therby I knowe, that yf 225 shepe dyd coste 45 l'i. euery shepe coste 4 s. *S.* This can I do, as you shall perceaue by this example: Yf 160 sowldyars do spende euery moneth 68 l'i. what spendeth eche man? Fyrst [1]bycause I can not diuide the 68 by 160, therfore I wyll turne the pou*n*des into pennes by multiplicacio*n*, so shall there be 16320 d'. Nowe muste I diuide this su*m*me by the 12 nomber of sowldyars, therfore I set the*m* in order, thus. Then begyn I at the hyghest place of the diuidente, sekynge my diuisor there, whiche I fynde ones, Therfore set I 1 in the nether lyne. *M.* Not in the nether line of the hole summe, but in the nether lyne of that worke, whiche is the thyrde lyne. *S.* So standeth it with reason. *M.* Then thus do they stande.[2] Then seke I agayne in the reste, how often I may fynde my diuisor, and I se that in the 300 I myghte fynde 100 thre tymes, but then the 60 wyll not be so often founde in 20, therfore I take 2 for my quotient: then take I 100 twyse from 300, and there resteth 100, out of whiche with the 20 (that maketh 120) I may take 60 also twyse, and then standeth the nombers thus, [3]where I haue sette the quotient 2 in the lowest lyne: So is euery sowldyars portion 102 d'. that is 8 s. 6 d'. *M.* But yet bycause you shall perceaue iustly the reason of Diuision, it shall be good that you do set your diuisor styll agaynst those nombres fro*m* whiche you do take it: as by this example I wyll declare. Yf ye purchace of 200 acres of ground dyd coste 290 l'i. what dyd one acre coste? Fyrst wyl I turne the poundes into pennes, so wyll there be 69600 d'. Then in settynge downe these nombers I shall do thus. Fyrst set the diuident on the ryghte hande as it oughte, and then [4]the diuisor on the lefte hande agaynst those nombers, from which I entende to take hym fyrst as here you se, wher I haue set the diuisor two lynes hygher the*n* is theyr owne place. *S.* This is lyke the order of diuision by the penne.

[1] 127 b.
[2] 128 a.
[3] 128 b.
[4] 129 a.

M. Truth you say, and nowe must I set yᵉ quotient of this worke in the thyrde lyne, for that is the lyne of vnities in respecte to the diuisor in this worke. Then I seke howe often the diuisor maye be founde in the diuident, *and* that I fynde 3 tymes, then set I 3 in the thyrde lyne for the quotient, and take awaye that 60000 fro*m* the diuident, and farther I do set the diuisor one line lower, as yow se here. [1]And then seke I how often the diuisor wyll be taken from the nomber agaynste it, whiche wyll be 4 tymes and 1 remaynynge. *S.* But what yf it chaunce that when the diuisor is so remoued, it can not be ones taken out of the diuident agaynste it? *M.* Then must the diuisor be set in an other line lower. *S.* So was it in diuision by the penne, and therfore was there a cypher set in the quotient: but howe shall that be noted here? *M.* Here nedeth no token, for the lynes do represente the places: onely loke that you set your quotient in that place which standeth for vnities in respecte of the diuisor: but now to returne to the example, I fynde the diuisor 4 tymes in the diuidente, and 1 remaynynge, for 4 tymes 2 make 8, which I take from 9, and there resteth 1, as this figure sheweth: and in the myddle space for the quotient I set 4 in the seconde lyne, whiche is in this worke the place of vnities.[2] Then remoue I yᵉ diuisor to the next lower line, and seke how often I may haue it in the dyuident, which I may do here 8 tymes iust, and nothynge remayne, as in this fourme, where you may se that the hole quoti- ent is 348 d', that is 29 s. wherby I knowe that so moche coste the purchace of one aker. *S.* Now resteth the profes of Multiplycatio*n*, and also of Diuisio*n*. *M.* Ther best profes are eche.[3] one by the other, Multyplication is proued by Diuision, and Diuision by Multiplyca- tion, as in the worke by the penne you learned. *S.* Yf that be all, you shall not nede to repete agayne that, yᵗ was sufficye*n*tly taughte all redye: and excepte you wyll teache me any other feate, here maye you make an ende of this arte I suppose. *M.* So wyll I do as touchynge hole nomber, and as for broken nomber, I wyll not trouble your wytte with it, tyll you haue practised this so well, yᵗ you be full perfecte, so that you nede not to doubte in any poynte that I haue taught you, and thenne maye I boldly enstructe you in yᵉ arte of fractions or broken no*m*ber, wherin I

[1] 129 *b.*

[2] 130 *a.*

[3] 130 *b.*

wyll also showe you the reasons of all that you haue nowe learned. But yet before I make an ende, I wyll showe you the order of co*m*men castyng, wher in are bothe pennes, shyllynges, and poundes, procedynge by no grounded reason, but onely by a receaued [1] fourme, and that dyuersly of dyuers men: for marcha*un*tes vse one fourme, and auditors an other: But fyrste for marchauntes fourme marke this example here, in which I haue expressed this summe 198 l'i.[2] 19 s. 11 d'. So that you maye se that the lowest lyne serueth for pe*n*nes, the next aboue for shyllynges, the thyr*d*e for poundes, and the fourth for scores of pou*n*des. And farther you maye se, that the space betwene pennes and shyllynges may receaue but one counter (as all other spaces lyke wayes do) and that one standeth in that place for 6 d'. Lyke wayes betwene the shyllynges *and* the pou*n*des, one cou*n*ter standeth for 10 s. And betwene the poundes and 20 l'i. one counter standeth for 10 poundes. But besyde those you maye see at the left syde of shyllynges, that one counter standeth alone, *and* betokeneth 5 s. [3]So agaynste the poundes, that one cou*n*ter standeth for 5 l'i. And agaynst the 20 poundes, the one counter standeth for 5 score pou*n*des, that is 100 l'i. so that euery syde counter is 5 tymes so moch as one of them agay*n*st whiche he standeth. Now for the accompt of auditors take this example. where I haue expressed y*e* same su*m*me 198 l'i. 19 s. 11 d'. But here you se the pe*n*nes stande toward y*e* ryght hande, and the other encreasynge orderly towarde the lefte hande. Agayne you maye se, that auditours wyll make 2 lynes (yea and more) for pennes, shyllynges, *and* all other valewes, yf theyr summes extende therto. Also you se, that they set one counter at the ryght ende of eche rowe, whiche so set there standeth for 5 of that roume: and on [4] the lefte corner of the rowe it sta*n*deth for 10, of y*e* same row. But now yf you wold adde other subtracte after any of both those sortes, yf you marke y*e* order of y*t* other feate which I taught you, you may easely do the same here without moch teachynge: for in Addition you must fyrst set downe one su*m*me and to the same set the other orderly, and lyke maner yf you haue many: but in Subtraction you must sette downe fyrst the greatest summe, and from it must you abate that other euery denomination from his dewe place. *S.* I do not doubte but with a

[2] 168 in original.

Auditors' Casting Counters.

lytell practise I shall attayne these bothe : but how shall I multiply and diuide after these fourmes ? *M.* You can not duely do none of both by these sortes, therfore in suche case, you must resort to your other artes. *S.* Syr, yet I se not by these sortes how to expresse hundreddes, yf they excede one hundred, nother yet thousandes. *M.* They that vse such accomptes that it excede 200 [1] in one summe, they sette no 5 at the lefte hande of the scores of poundes, but they set all the hundredes in an other farther rowe *and* 500 at the lefte hand therof, and the thousandes they set in a farther rowe yet, *and* at the lefte syde therof they sette the 5000, and in the space ouer they sette the 10000, and in a hygher rowe 20000, whiche all I haue expressed in this example, which is 97869l'i. 12s. 9d' ob. q. for I had not told you before where, nother how you shuld set downe farthynges, which (as you se here) must be set in a voyde space sydelynge beneth the pennes : for q one counter : for ob. 2 counters : for ob. q. 3 counters : *and* more there can not be, for 4 farthynges [2] do make 1 d'. which must be set in his dewe place. And yf you desyre ye same summe after audytors maner, lo here it is.

[1] 132 *b.*

[2] 133 *a.*

But in this thyng, you shall take this for suffycyent, and the reste you shall obserue as you maye se by the working of eche sorte : for the dyuers wittes of men haue inuented dyuers and sundry wayes almost vnnumerable. But one feate I shall teache you, whiche not only for the straungenes and secretnes is moche pleasaunt, but also for the good commoditie of it ryghte worthy to be well marked. This feate hath ben vsed aboue 2000 yeares at the leaste, and yet was it neuer comenly knowen, especyally in Englysshe it was neuer taughte yet. This is the arte of nombrynge on the hand, with diuers gestures of the fyngers, expressynge any summe conceaued in the [3] mynde. And fyrst to begynne, yf you wyll expresse any summe vnder 100, you shall expresse it with your lefte hande : and from 100 vnto 10000, you shall expresse it with your ryght hande, as here orderly by this table folowynge you may perceaue.

[3] 133 *b.*

¶ Here foloweth the table
of the arte of the
hande

NOMBRYNGE.

The arte of nombrynge by the hande.

134

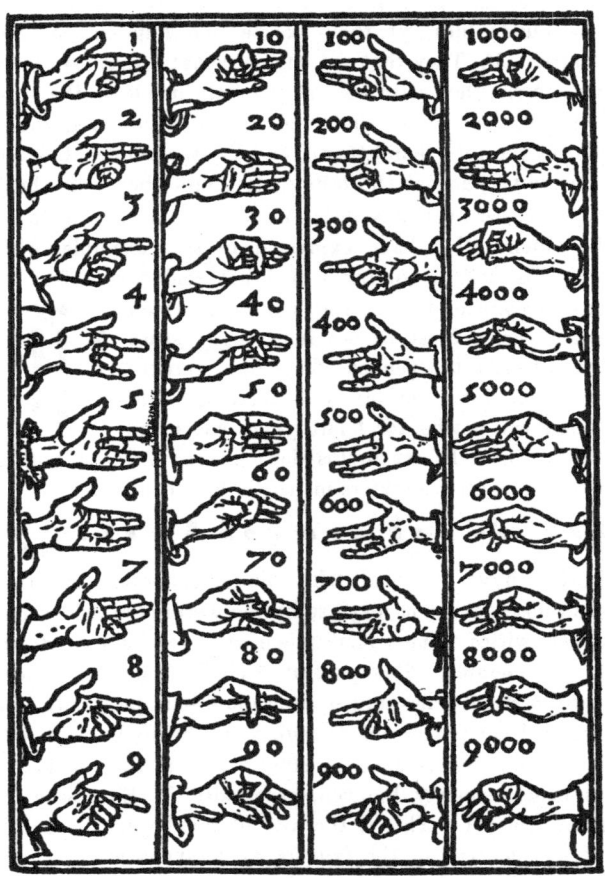

¹ 134 b. ¹ ¹In which as you may se 1 is expressed by yͤ lyttle fynger of yͤ
² lefte hande closely and harde croked. *[2 is declared by lyke bowynge of the weddynge fynger (whiche is the nexte to the lyttell
³ fynger) together with the lytell fynger. [3 is signified by the
myddle fynger bowed in lyke maner, with those other two. [4 is
declared by the bowyng of the myddle fynger and the rynge

* Bracket ([) denotes new paragraph in original.

Digital Signs of Numbers.

fynger, or weddynge fynger, with the other all stretched forth.
[5 is represented by the myddle fynger onely bowed. [And 6 by
the weddynge fynger only crooked: and this you may marke in
these a certayne order. But now 7, 8, and 9, are expressed with
the bowynge of the same fyngers as are 1, 2, and 3, but after an
other fourme. [For 7 is declared by the bowynge of the lytell
fynger, as is 1, saue that for 1 the fynger is clasped in, harde *and*
rounde, but for to expresse 7, you shall bowe the myddle ioynte
of the lytell fynger only, and holde the other ioyntes streyght.
S. Yf you wyll geue me leue to expresse it after my rude maner,
thus I vnderstand your meanyng : that 1 is expressed by crookynge
in the lyttell fynger lyke the head of a bysshoppes bagle : and 7 is
declared by the same fynger bowed lyke a gybbet. *M.* So I
perceaue, you vnderstande it. [Then to expresse 8, you shall bowe
after the same maner both the lyttell fynger and the rynge fynger.
[And yf you bowe lyke wayes with them the myddle fynger, then
doth it betoken 9. [Now to expresse 10, you shall bowe your
fore fynger rounde, and set the ende of it on the hyghest ioynte of
the thombe. [And for to expresse 20, you must set your fyngers
streyght, and the ende of your thombe to the partition of the fore
moste and myddle fynger. [30 is represented by the ioynynge
together of ye headdes of the foremost fynger and the thombe.
[40 is declared by settynge of the thombe crossewayes on the foremost fynger. [50 is signified by ryght stretchyng forth of the
fyngers ioyntly, and applyenge of the thombes ende to the partition
of the myddle fynger *and* the rynge fynger, or weddynge fynger.
[60 is formed by bendynge of the thombe croked and crossynge it
with the fore fynger. [70 is expressed by the bowynge of the
foremost fynger, and settynge the ende of the thombe betweene the
2 foremost or hyghest ioyntes of it. [80 is expressed by settynge
of the foremost fynger crossewayes on the thombe, so that 80
dyffereth thus from 40, that for 80 the forefynger is set crosse on
the thombe, and for 40 the thombe is set crosse ouer ye forefinger.
[90 is signified, by bendynge the fore fynger, and settyng the ende
of it in the innermost ioynte of ye thombe, that is euen at the foote
of it. And thus are all the nombers ended vnder 100. *S.* In
dede these be all the nombers from 1 to 10, *and* then all the
tenthes within 100, but this teacyed me not how to expresse 11,
12, 13, *etc.* 21, 22, 23, *etc.* and such lyke. *M.* You can lytell
vnderstande, yf you can not do that without teachynge: what is

11? is it not 10 and 1? then expresse 10 as you were taught, and 1 also, and that is 11 : and for 12 expresse 10 and 2 : for 23 set 20 and 3 : and so for 68 you muste make 60 and there to 8 : and so ₁₀₀ of all other sortes. [But now yf you wolde represente 100 other any nomber aboue it, you muste do that with the ryghte hande, after this maner. [You must expresse 100 in the ryght hand, with the lytell fynger so bowed as you dyd expresse 1 in the left hand.

¹ 136 b.   ¹[And as you expressed 2 in the lefte hande, the same fasshyon ₂₀₀ in the ryght hande doth declare 200.

₃₀₀   The fourme of 3 in the ryght hand standeth for 300.
₄₀₀   The fourme of 4, for 400.
₅₀₀   Lykewayes the fourme of 5, for 500.
₆₀₀   The fourme of 6, for 600. And to be shorte : loke how you did expresse single vnities and tenthes in the lefte hande, so must you expresse vnities *and* tenthes of hundredes, in the ryghte hande.
₉₀₀ *S.* I vnderstande you thus : that yf I wold represent 900, I must so fourme the fyngers of my ryghte hande, as I shuld do in my left hand to expresse 9, And as in my lefte hand I expressed ₁₀₀₀ 10, so in my ryght hande must I expresse 1000.

And so the fourme of euery tenthe in the lefte hande serueth to expresse lyke no*m*ber of thousa*n*des, so ye fourme of 40 standeth ₄₀₀₀ for 4000.

₈₀₀₀   The fourme of 80 for 8000.

² 137 a.   ²And the fourme of 90 (whiche is
₉₀₀₀ the greatest) for 9000, and aboue that
I can not expresse any nomber. *M.*
No not with one fynger : how be it,
wi*th* dyuers fyngers you maye expresse
9999, and all at one tyme, and that lac
keth but 1 of 10000. So that vnder
10000 you may by your fyngers expresse any summe. And this shal suf-
fyce for Numeration on the fyngers.
And as for Addition, Subtraction,
Multiplicatio*n*, and Diuision (which
yet were neuer taught by any man as
farre as I do knowe) I wyll enstruct
you after the treatyse of fractions.
And now for this tyme fare well,

and loke that you cease not to
practyse that you haue learned. *S.* Syr, with moste
harty mynde I thanke
you, bothe for your
good learnyng, *and*
also your good
counsel, which
(god wyllyng) I truste to folow.

Finis.

APPENDIX I.

A Treatise on the Numeration of Algorism.

[*From a MS. of the 14th Century.*]

To alle suche even nombrys the most have cifrys as to ten. twenty. thirtty. an hundred. an thousand and suche other. but ye schal vnderstonde that a cifre tokeneth nothinge but he maketh other the more significatyf that comith after hym. Also ye schal vnderstonde that in nombrys composyt and in alle other nombrys that ben of diverse figurys ye schal begynne in the ritht syde and to rekene backwarde and so he schal be wryte as thus—1000. the sifre in the ritht side was first wryte and yit he tokeneth nothinge to the secunde no the thridde but thei maken that figure of 1 the more signyficatyf that comith after hem by as moche as he born oute of his first place where he schuld yf he stode ther .tokene but one. And there he stondith nowe in the ferye place he tokeneth a thousand as by this rewle. In the first place he tokeneth but hymself. In the secunde place he tokeneth ten tymes hymself. In the thridde place he tokeneth an hundred tymes himself. In the ferye he tokeneth a thousand tymes himself. In the fyftye place he tokeneth ten thousand tymes himself. In the sexte place he tokeneth an hundred thousand tymes hymself. In the seventh place he tokeneth ten hundred thousand tymes hymself, &c. And ye schal vnderstond that this worde nombre is partyd into thre partyes. Somme is callyd nombre of digitys for alle ben digitys that ben withine ten as ix, viii, vii, vi, v, iv, iii, ii, i. Articules ben alle thei that mow be devyded into nombrys of ten as xx, xxx, xl, and suche other. Composittys be alle nombrys that ben componyd of a digyt and of an articule as fourtene fyftene thrittene and suche other. Fourtene is componyd of four that is a digyt

Numeration.

and of ten that is an articule. Fyftene is componyd of fyve that is a digyt and of ten that is an articule and so of others . . . . . . But as to this rewle. In the firste place he tokeneth but himself that is to say he tokeneth but that and no more. If that he stonde in the secunde place he tokeneth ten tymes himself as this figure 2 here 21. this is oon and twenty. This figure 2 stondith in the secunde place and therfor he tokeneth ten tymes himself and ten tymes 2 is twenty and so forye of every figure and he stonde after another toward the lest syde he schal tokene ten tymes as moche more as he schuld token and he stode in that place ther that the figure afore him stondeth: lo an example as thus 9634. This figure of foure that hath this schape 4 tokeneth but himself for he stondeth in the first place. The figure of thre that hath this schape 3 tokeneth ten tyme himself for he stondeth in the secunde place and that is thritti. The figure of sexe that hath this schape 6 tokeneth ten tyme more than he schuld and he stode in the place yer the figure of thre stondeth for ther he schuld tokene but sexty. And now he tokeneth ten tymes that is sexe hundrid. The figure of nyne that hath this schape 9 tokeneth ten tymes more than he schulde and he stode in the place ther the figure of 6 stondeth inne for thanne he schuld tokene but nyne hundryd. And in the place that he stondeth inne nowe he tokeneth nine thousand. Alle the hole nombre of these foure figurys. Nine thousand sexe hundrid and foure and thritti.

APPENDIX II.

Carmen de Algorismo.

[*From a B.M. MS., 8 C. iv., with additions from 12 E. 1 & Eg. 2622.*]

Hec algorismus ars presens dicitur[1] ; in qua
Talibus Indorum[2] fruimur bis quinque figuris.
  0. 9. 8. 7. 6. 5. 4. 3. 2. 1.
Prima significat unum : duo vero secunda :   4
Tercia significat tria : sic procede sinistre
Donec ad extremam venies, qua cifra vocatur ;
[3][Que nil significat ; dat significare sequenti.]
Quelibet illarum si primo limite ponas,   8
Simpliciter se significat : si vero secundo,
Se decies : sursum procedas multiplicando.[4]
[Namque figura sequens quevis signat decies plus,
Ipsa locata loco quam significet pereunte :   12
Nam precedentes plus ultima significabit.]

[5]Post predicta scias quod tres breuiter numerorum
Distincte species sunt ; nam quidam digiti sunt ;
Articuli quidam ; quidam quoque compositi sunt.   16
[Sunt digiti numeri qui citra denarium sunt ;
Articuli decupli degitorum ; compositi sunt
Illi qui constant ex articulis digitisque.]
Ergo, proposito numero tibi scribere, primo   20
Respicias quis sit numerus ; quia si digitus sit,
[5][Una figura satis sibi ; sed si compositus sit,]
Primo scribe loco digitum post articulum fac
Articulus si sit, cifram post articulum sit,   24
[Articulum vero reliquenti in scribe figure.]

 [1] "Hec præsens ars dicitur algorismus ab Algore rege ejus inventore, vel dicitur ab *algos* quod est ars, et *rodos* quod est numerus ; quæ est ars numerorum vel numerandi, ad quam artem bene sciendum inveniebantur apud Indos bis quinque (id est decem) figuræ."—*Comment. Thomæ de Novo-Mercatu.* MS. Bib. Reg. Mus. Brit. 12 E. 1.
 [2] "Hæ necessariæ figuræ sunt Indorum characteros." *MS. de numeratione.* Bib. Sloan. Mus. Brit. 513, fol. 58. "Cum vidissem Yndos constituisse ix literas in universo numero suo propter dispositionem suam quam posuerunt, volui patefacere de opere quod sit per eas aliquidque esset levius discentibus, si Deus voluerit. Si autem Indi hoc voluerunt et intentio illorum nihil novem literis fuit, causa que mihi potuit. Deus direxit me ad hoc. Si vero alia dicam preter eam quam ego exposui, hoc fecerunt per hoc quod ego exposui, eadem tam certissime et absque ulla dubitatione poterit inveniri. Levitasque patebit aspicientibus et discentibus." MS. U. L. C., li. vi. 5, f. 102.
 [3] From Eg. 2622.
 [4] 8 C. iv. inserts Nullum cipa significat : dat significare sequenti.
 [5] From 12 E. 1.

Quolibet in numero, si par sit prima figura,
Par erit et totum, quicquid sibi continetur;
Impar si fuerit, totum sibi fiet et impar.   28

Septem[1] sunt partes, non plures, istius artis;
Addere, subtrahere, duplare, dimidiare;
Sexta est diuidere, set quinta est multiplicare;
Radicem extrahere pars septima dicitur esse.   32
Subtrahis aut addis a dextris vel mediabis;
A leua dupla, diuide, multiplicaque;
Extrahe radicem semper sub parte sinistra.

Addere si numero numerum vis, ordine tali   36 Addition.
Incipe; scribe duas primo series numerorum
Prima sub prima recte ponendo figuram,
Et sic de reliquis facias, si sint tibi plures.
Inde duas adde primas hac condicione;   40
Si digitus crescat ex addicione priorum,
Primo scribe loco digitum, quicunque sit ille;
Si sit compositus, in limite scribe sequenti
Articulum, primo digitum; quia sic iubet ordo.   44
Articulus si sit, in primo limite cifram,
Articulum vero reliquis inscribe figuris;
Vel per se scribas si nulla figura sequatur.
Si tibi cifra superueniens occurrerit, illam   48
Deme suppositam; post illic scribe figuram:
Postea procedas reliquas addendo figuras.

A numero numerum si sit tibi demere cura,   Subtraction.
Scribe figurarum series, vt in addicione;   52
Maiori numero numerum suppone minorem,
Siue pari numero supponatur numerus par.
Postea si possis a prima subtrahe primam,
Scribens quod remanet, cifram si nil remanebit.   56
Set si non possis a prima demere primam;
Procedens, vnum de limite deme sequenti;

[1] En argorisme devon prendre
Vii especes . . . .
Adision subtracion
Doubloison mediacion
Monteploie et division
Et de radix enstracion
A chez vii especes savoir
Doit chascun en memoire avoir
Letres qui figures sont dites
Et qui excellens sont ecrites.—MS. *Seld. Arch.* B. 26.

	Et demptum pro denario reputabis ab illo,	
	Subtrahe totaliter numerum quem proposuisti.	60
	Quo facto, scribe supra quicquit remanebit,	
	Facque novenarios de cifris, cum remanebis,	
	Occurrant si forte cifre, dum demseris vnum;	
	Postea procedas reliquas demendo figuras.	64
Proof.	[1] [Si subtracio sit bene facta probare valebis,	
	Quas subtraxisti primas addendo figuras.	
	Nam, subtractio si bene sit, primas retinebis,	
	Et subtractio facta tibi probat additionem.]	68
Duplation.	Si vis duplare numerum, sic incipe; solam	
	Scribe figurarum seriem, quamcumque voles que	
	Postea procedas primam duplando figuram;	
	Inde quod excrescet, scribens, vbi iusserit ordo,	72
	Juxta precepta que dantur in addicione.	
	Nam si sit digitus, in primo limite scribe;	
	Articulus si sit, in primo limite cifram,	
	Articulum vero reliquis inscribe figuris;	76
	Vel per se scribas, si nulla figura sequatur:	
	Compositus si sit, in limite scribe sequenti	
	Articulum primo, digitum; quia sic jubet ordo:	
	Et sic de reliquis facias, si sint tibi plures.	80
	[1] [Si super extremam nota sit, monadem dat eidem,	
	Quod tibi contingit, si primo dimidiabis.]	
Mediation.	Incipe sic, si vis aliquem numerum mediare:	
	Scribe figurarum seriem solam, velud ante;	84
	Postea procedens medias, et prima figura	
	Si par aut impar videas; quia si fuerit par,	
	Dimidiabis eam, scribens quicquit remanebit;	
	Impar si fuerit, vnum demas, mediare,	88
	Nonne presumas, sed quod superest mediabis;	
	Inde super tractum, fac demptum quod notat unum;	
	Si monos, dele; sit ibi cifra post nota supra.	
	Postea procedas hac condicione secunda:[2]	92
	Impar[3] si fuerit hic vnum deme priori,	
	Inscribens quinque, nam denos significabit	
	Monos prædictam: si vero secunda dat vnam,	
	Illa deleta, scribatur cifra; priori	96

[1] From 12 E. 1.
[2] 8 C. iv. inserts Atque figura prior nuper fuerit mediando.
[3] *I. e.* figura secundo loco posita.

Tradendo quinque pro denario mediato ;
Nec cifra scribatur, nisi inde figura sequatur :
Postea procdeas reliquas mediando figuras,
Quin supra docui, si sint tibi mille figure. 100
[1] [Si mediatio sit bene facta probare valebis,
Duplando numerum quem primo dimidiasti.]

Si tu per numerum numerum vis multiplicare,
Scribe duas, quascunque volis, series numerorum ; 104
Ordo tamen seruetur vt vltima multiplicandi
Ponatur super anteriorem multiplicantis ;
[2] [A leua relique sint scripte multiplicantes.]
In digitum cures digitum si ducere, major 108
Per quantes distat a denis respice, debes
Namque suo decuplo tociens delere minorem ;
Sicque tibi numerus veniens exinde patebit.
Postea procedas postremam multiplicando, 112
Juste multiplicans per cunctas inferiores,
Condicione tamen tali ; quod multiplicantis
Scribas in capite, quicquid processerit inde ;
Set postquam fuerit hec multiplicata, figure 116
Anteriorentur seriei multiplicantis ;
Et sic multiplica, velut istam multiplicasti,
Qui sequitur numerum scriptum quicunque figuris.
Set cum multiplicas, primo sic est operandum, 120
Si dabit articulum tibi multiplicacio solum ;
Proposita cifra, summam transferre memento.
Sin autem digitus excrescerit articulusque,
Articulus supraposito digito salit ultra ; 124
Si digitus tamen, ponas illum super ipsam,
Subdita multiplicans hanc que super incidit illi
Delet eam penitus, scribens quod provenit inde ;
Sed si multiplices illam posite super ipsam, 128
Adiungens numerum quem prebet ductus earum ;
Si supraimpositam cifra debet multiplicare,
Prorsus eam delet, scribi que loco cifra debet,
[2] [Si cifra multiplicat aliam positam super ipsam, 132
Sitque locus supra vacuus super hanc cifra fiet ;]

[1] So 12 E. 1 ; 8 C. iv. inserts—
    Si super extremam nota sit monades dat eidem
    Quod contingat cum primo dimiabis
      Atque figura prior nuper fuerit mediando.
[2] 12 E. 1 inserts.

Mental Multiplication.

Si supra fuerit cifra semper pretereunda est;
Si dubites, an sit bene multiplicando secunda,
Diuide totalem numerum per multiplicantem,   136
Et reddet numerus emergens inde priorem.
[1][Per numerum si vis numerum quoque multiplicare
Tantum per normas subtiles absque figuris
Has normas poteris per versus scire sequentes.   140
Si tu per digitum digitum quilibet multiplicabis
Regula precedens dat qualiter est operandum
Articulum si per reliquum vis multiplicare
In proprium digitum debebit uterque resolvi   144
Articulus digitos post per se multiplicantes
Ex digitis quociens teneret multiplicatum
Articuli faciunt tot centum multiplicati.
Articulum digito si multiplicamus oportet   148
Articulum digitum sumi quo multiplicare
Debemus reliquum quod multiplicaris ab illis
Per reliquo decuplum sic omne latere nequibit
In numerum mixtum digitum si ducere cures   152
Articulus mixti sumatur deinde resolvas
In digitum post hec fac ita de digitis nec
Articulusque docet excrescens in detinendo
In digitum mixti post ducas multiplicantem   156
De digitis ut norma docet sit juncta secundo
Multiplica summam et postea summa patebit
Junctus in articulum purum articulumque
[2][Articulum purum comittes articulum que]   160
Mixti pro digitis post fiat et articulus vt
Norma jubet retinendo quod egreditur ab illis
Articuli digitum post iu digitum mixti duc
Regula de digitis ut percipit articulusque   164
Ex quibus excrescens summe tu junge priori
Sic manifesta cito fiet tibi summa petita.
Compositum numerum mixto sic multiplicabis
Vndecies tredecem sic est ex hiis operandum   168
In reliquum primum demum duc post in eundem
Unum post deinde duc in tercia deinde per unum
Multiplices tercia demum tunc omnia multiplicata
In summa duces quam que fuerit te dices   172

---

[1] 12 E. 1 inserts to l. 174.   [2] 12 E. 1 omits, Eg. 2622 inserts.

Division, Square Numbers.

Hic ut hic mixtus intentus est operandum
Multiplicandorum de normis sufficiunt hec.]
Si vis dividere numerum, sic incipe primo ;   *Division.*
Scribe duas, quascunque voles, series numerorum ;   176
Majori numero numerum suppone minorem,
[1][Nam docet ut major teneat bis terve minorem ;]
Et sub supprima supprimam pone figuram,
Sic reliquis reliquas a dextra parte locabis ;   180
Postea de prima primam sub parte sinistra
Subtrahe, si possis, quociens potes adminus istud,
Scribens quod remanet sub tali conditione ;
Ut totiens demas demendas a remanente,   184
Que serie recte ponentur in anteriori,
Unica si, tantum sit ibi decet operari ;
Set si non possis a prima demere primam,
Procedas, et eam numero suppone sequenti ;   188
Hanc uno retrahendo gradu quo comites retrahantur,
Et, quotiens poteris, ab eadem deme priorem,
Ut totiens demas demendas a remanenti,
Nec plus quam novies quicquam tibi demere debes,   192
Nascitur hinc numerus quociens supraque sequentem
Hunc primo scribas, retrahas exinde figuras,
Dum fuerit major supra positus inferiori,
Et rursum fiat divisio more priori ;   196
Et numerum quotiens supra scribas pereunti,
Si fiat saliens retrahendo, cifra locetur,
Et pereat numero quotiens, proponas eidem
Cifram, ne numerum pereat vis, dum locus illic   200
Restat, et expletis divisio non valet ultra :
Dum fuerit numerus numerorum inferiore seorsum
Illum servabis ; hinc multiplicando probabis,
Si bene fecisti, divisor multiplicetur   204 *Proof.*
Per numerum quotiens ; cum multiplicaveris, adde
Totali summæ, quod servatum fuit ante,
Reddeturque tibi numerus quem proposuisti ;
Et si nil remanet, hunc multiplicando reddet,   208
Cum ducis numerum per se, qui provenit inde   *Square Numbers.*
Sit tibi quadratus, ductus radix erit hujus,
Nec numeros omnes quadratos dicere debes,
Est autem omnis numerus radix alicujus.   212

[1] 12 E. 1 inserts.

Square Root.

Quando voles numeri radicem querere, scribi
Debet; inde notes si sit locus ulterius impar,
Estque figura loco talis scribenda sub illo,
Que, per se dicta, numerum tibi destruat illum, 216
Vel quantum poterit ex inde delebis eandem;
Vel retrahendo duples retrahens duplando sub ista
Que primo sequitur, duplicatur per duplacationem,
Post per se minuens pro posse quod est minuendum. 220
[1] Post his propones digitum, qui, more priori
Per precedentes, post per se multiplicatus,
Destruat in quantum poterit numerum remanentem,
Et sic procedens retrahens duplando figuram, 224
Preponendo novam donec totum peragatur,
Subdupla propriis servare docetque duplatis;
Si det compositum numerum duplacio, debet
Inscribi digitus a parte dextra parte propinqua, 228
Articulusque loco quo non duplicata resessit;
Si dabit articulum, sit cifra loco pereunte
Articulusque locum tenet unum, de duplicata resessit;
Si donet digitum, sub prima pone sequente, 232
Si supraposita fuerit duplicata figura
Major proponi debet tantummodo cifra,
Has retrahens solito propones more figuram,
Usque sub extrema ita fac retrahendo figuras, 236
Si totum deles numerum quem proposuisti,
Quadratus fuerit, de dupla quod duplicasti,
Sicque tibi radix illius certa patebit,
Si de duplatis fit juncta supprima figura; 240
Radicem per se multiplices habeasque
Primo propositum, bene te fecisse probasti;
Non est quadratus, si quis restat, sed habentur
Radix quadrati qui stat major sub eadem; 244
Vel quicquid remanet tabula servare memento;
Hoc casu radix per se quoque multiplicetur,
Vel sic quadratus sub primo major habetur,
Hinc addas remanens, et prius debes haberi; 248
Si locus extremus fuerit par, scribe figuram
Sub pereunte loco per quam debes operari,
Que quantum poterit supprimas destruat ambas,

[1] 8 C. iv. inserts—
Hinc illam dele duplans sub ei psalliendo
Que sequitur retrahens quicquid fuerit duplicatum.

Vel penitus legem teneas operando priorem, 252
Si suppositum digitus suo fine repertus,
Omnino delet illic scribi cifra debet,
A leva si qua sit ei sociata figura;
Si cifre remanent in fine pares decet harum 256
Radices, numero mediam proponere partem,
Tali quesita radix patet arte reperta.
Per numerum recte si nosti multiplicare
Ejus quadratum, numerus qui pervenit inde 260
Dicetur cubicus; primus radix erit ejus;
Nec numeros omnes cubicatos dicere debes,
Est autem omnis numerus radix alicujus;
Si curas cubici radicem quærere, primo 264
Inscriptum numerum distinguere per loca debes;
Que tibi mille notant a mille notante suprema
Initiam, summa operandi parte sinistra,
Illic sub scribas digitum, qui multiplicatus 268
In semet cubice suprapositum sibi perdat,
Et si quid fuerit adjunctum parte sinistra
Si non omnino, quantum poteris minuendo,
Hinc triplans retrahe saltum, faciendo sub illa 272
Que manet a digito deleto terna, figuram
Illi propones que sub triplo asocietur,
Ut cum subtriplo per eam tripla multiplicatur;
Hinc per eam solam productum multiplicabis, 276
Postea totalem numerum, qui provenit inde
A suprapositis respectu tolle triplate
Addita supprimo cubice tunc multiplicetur,
Respectu cujus, numerus qui progredietur 280
Ex cubito ductu, supra omnes adimetur;
Tunc ipsam delens triples saltum faciendo,
Semper sub ternas, retrahens alias triplicatas
Ex hinc triplatis aliam propone figuram, 284
Que per triplatas ducatur more priori;
Primo sub triplis sibi junctis, postea per se,
In numerum ducta, productum de triplicatis:
Utque prius dixi numerus qui provenit inde 288
A suprapositis has respiciendo trahatur,
Huic cubice ductum sub primo multiplicabis,
Respectumque sui, removebis de remanenti,
Et sic procedas retrahendo triplando figuram. 292

Cube Root.

Et proponendo nonam, donec totum peragatur,
Subtripla sub propriis servare decet triplicatis ;
Si nil in fine remanet, numerus datus ante
Est cubicus ; cubicam radicem sub tripla prebent,   296
Cum digito juncto quem supprimo posuisti,
Hec cubice ducta, numerum reddant tibi primum.
Si quid erit remanens non est cubicus, sed habetur
Major sub primo qui stat radix cubicam,   300·
Servari debet quicquid radice remansit,
Extracto numero, decet hec addi cubicato.
Quo facto, numerus reddi debet tibi primus.
Nam debes per se radicem multiplicare   304
Ex hinc in numerum duces, qui provenit inde
Sub primo cubicus major sic invenietur ;
Illi jungatur remanens, et primus habetur,.
Si per triplatum numerum nequeas operari ;   308
Cifram propones, nil vero per hanc operare
Set retrahens illam cum saltu deinde triplata,
Propones illi digitum sub lege priori,
Cumque cifram retrahas saliendo, non triplicabis,   312
Namque nihil cifre triplacio dicitur esse ;
At tu cum cifram protraxeris aut triplicata,
Hanc cum subtriplo semper servare memento :
Si det compositum, digiti triplacio debet   316
Illius scribi, digitus saliendo sub ipsam ;
Digito deleto, que terna dicitur esse ;
Jungitur articulus cum triplata pereunte,
Set facit hunc scribi per se triplacio prima,   320
Que si det digitum per se scribi facit illum ;
Consumpto numero, si sole fuit tibi cifre
Triplato, propone cifram saltum faciendo,
Cumque cifram retrahe triplam, scribendo figuram,   324
Preponas cifre, sic procedens operare,
Si tres vel duo serie in sint, pone sub yma,
A dextris digitum servando prius documentum.
Si sit continua progressio terminus nuper   328
Per majus medium totalem multiplicato ;
Si par, per medium tunc multiplicato sequentem.
Set si continua non sit progressio finis :
Impar, tunc majus medium si multiplicabis,   332
Si par per medium sibi multiplicato propinquum.   333·

INDEX OF TECHNICAL TERMS[1]

algorisme, 33/12 ; **algorym, augrym,** 3/3 ; the art of computing, using the so-called Arabic numerals.

The word in its various forms is derived from the Arabic *al-Khowarazmi* (i. e. the native of Khwarazm (Khiva)). This was the surname of Ja'far Mohammad ben Musa, who wrote a treatise early in the 9th century (see p. xiv).

The form *algorithm* is also found, being suggested by a supposed derivation from the Greek ἀριθμός (number).

antery, 24/11 ; to move figures to the right of the position in which they are first written. This operation is performed repeatedly upon the multiplier in multiplication, and upon certain figures which arise in the process of root extraction.

anterioracioun, 50/5 ; the operation of moving figures to the right.

article, 34/23 ; **articul,** 5/31 ; **articuls,** 9/36, 29/7, 8 ; a number divisible by ten without remainder.

cast, 8/12 ; to add one number to another.

'Addition is a *casting* together of two numbers into one number,' 8/10.

cifre, 4/1 ; the name of the figure 0. The word is derived from the Arabic *sifr* = empty, nothing. Hence *zero*.

A cipher is the symbol of the absence of number or of zero quantity. It may be used alone or in conjunction with digits or other ciphers, and in the latter case, according to the position which it occupies relative to the other figures, indicates the absence of units, or tens, or hundreds, etc. The great superiority of the Arabic to all other systems of notation resides in the employment of this symbol. When the cipher is not used, the place value of digits has to be indicated by writing them in assigned rows or columns. Ciphers, however, may be interpolated amongst the significant figures used, and as they sufficiently indicate the positions of the empty rows or columns, the latter need not be indicated in any other way. The practical performance of calculations is thus enormously facilitated (see p. xvi).

componede, 33/24 ; **composyt,** 5/35 ; with reference to numbers, one compounded of a multiple of ten and a digit.

conuertide = conversely, 46/29, 47/9.

cubicede, 50/13 ; **to be c.,** to have its cube root found.

[1] This Index has been kindly prepared by Professor J. B. Dale, of King's College, University of London, and the best thanks of the Society are due to him for his valuable contribution.

cubike nombre, 47/8; a number formed by multiplying a given number twice by itself, e.g. $27 = 3 \times 3 \times 3$. Now called simply a cube.

decuple, 22/12; the product of a number by ten. Tenfold.

departys = divides, 5/29.

digit, 5/30; **digitalle,** 33/24; a number less than ten, represented by one of the nine Arabic numerals.

dimydicion, 7/23; the operation of dividing a number by two. Halving.

duccioun, multiplication, 43/9.

duplacion, 7/23, 14/15; the operation of multiplying a number by two. Doubling.

i-mediet = halved, 19/23.

intercise = broken, 46/2; intercise Progression is the name given to either of the Progressions 1, 3, 5, 7, etc.; 2, 4, 6, 8, etc., in which the common difference is 2.

lede into, multiply by, 47/18.

lyneal nombre, 46/14; a number such as that which expresses the measure of the length of a line, and therefore is not *necessarily* the product of two or more numbers (*vide* Superficial, Solid). This appears to be the meaning of the phrase as used in *The Art of Nombryng*. It is possible that the numbers so designated are the prime numbers, that is, numbers not divisible by any other number except themselves and unity, but it is not clear that this limitation is intended.

mediacioun, 16/36, 38/16; dividing by two (see also **dimydicion**).

medlede nombre, 34/1; a number formed of a multiple of ten and a digit (*vide* componede, composyt).

medye, 17/8, to halve; **mediete,** halved, 17/30; **ymedit,** 20/9.

naturelle progressioun, 45/22; the series of numbers 1, 2, 3, etc.

produccioun, multiplication, 50/11.

quadrat nombre, 46/12; a number formed by multiplying a given number by itself, e.g. $9 = 3 \times 3$, a square.

rote, 7/25; **roote,** 47/11; root. The roots of squares and cubes are the numbers from which the squares and cubes are derived by multiplication into themselves.

significatyf, significant, 5/14. The significant figures of a number are, strictly speaking, those other than zero, e.g. in 3 6 5 0 4 0 0, the significant figures are 3, 6, 5, 4. Modern usage, however, regards all figures between the two extreme significant figures as significant, even when some are zero. Thus, in the above example, 3 6 5 0 4 are considered significant.

solide nombre, 46/37; a number which is the product of three other numbers, e.g. $66 = 11 \times 2 \times 3$.

superficial nombre, 46/18; a number which is the product of two other numbers, e.g. $6 = 2 \times 3$.

ternary, consisting of three digits, 51/7.

vnder double, a digit which has been doubled, 48/3.

vnder-trebille, a digit which has been trebled, 49/28; **vnder-triplat,** 49/39.

w, a symbol used to denote half a unit, 17/33.

GLOSSARY

ablacioun, taking away, 36/21
addyst, haddest, 10/37
agregacioun, addition, 45/22. (First example in N.E.D., 1547.)
a-ȝenenes, against, 23/10
allgate, always, 8/39
als, as, 22/24
and, if, 29/8; &, 4/27; & yf, 20/7
a-nendes, towards, 23/15
aproprede, appropriated, 34/27
apwereth, appears, 61/8
a-risyȝt, arises, 14/24
a-rowe, in a row, 29/10
arsemetrike, arithmetic, 33/1
ayene, again, 45/15

bagle, crozier, 67/12
bordure = ordure, row, 43/30
borro, *inf.* borrow, 11/38; *imp. s.* borewe, 12/20; *pp.* berwed, 12/15; borred, 12/19
boue, above, 42/34

caputule, chapter, 7/26
certayn, assuredly, 18/34
clepede, called, 47/7
competently, conveniently, 35/8
compt, count, 47/29
contynes, contains, 21/12; *pp.* contenythe, 38/39
craft, art, 3/4

distingue, divide, 51/5

egalle, equal, 45/21
excep, except, 5/16
exclusede, excluded, 34/37
excressent, resulting, 35/16
exeant, resulting, 43/26
expone, expound, 3/23

ferye = ferþe, fourth, 70/12
figure = figures, 5/1
for-by, past, 11/21
fors; no f., no matter, 22/24
forseth, matters, 53/30
forye = forþe, forth, 71/8
fyftye = fyftþe, fifth, 70/16

grewe, Greek, 33/13

haluendel, half, 16/16; haldel, 19/4; *pl.* haluedels, 16/16
hayst, hast, 17/3, 32
hast, haste, 22/25
heer, higher, 9/35
here, their, 7/26
here-a-fore, heretofore, 13/7
heyth, was called, 3/5
hole, whole, 4/39; holle, 17/1; hoole, of three dimensions, 46/15
holdyþe, holds good, 30/5
how be it that, although, 44/4

lede = lete, let, 8/37
lene, lend, 12/39
lest, least, 43/27
lest = left, 71/9
leue, leave, 6/5; *pr.* 3 *s.* leues, remains, 11/19; leus, 11/28; *pp.* laft, left, 19/24
lewder, more ignorant, 3/3
lust, desirest to, 45/13
lyȝt, easy, 15/31
lymytes, limits, 34/18; lynes, 34/12; lynees, 34/17; Lat. limes, *pl.* limites.

maystery, achievement; no m., no achievement, i. e. easy, 19/10
me, *indef. pron.* one, 42/1
mo, more, 9/16

Glossary.

moder = more (Lat. majorem), 43/22
most, must, 30/3
multipliede, to be m. = multiplying, 40/9
mynvtes, the sixty parts into which a unit is divided, 38/25
myse-wroȝt, mis-wrought, 14/11

nether, nor, 34/25
nex, next, 19/9
noȝt, nought, 5/7
note, not, 30/5

oo, one, 42/20; o, 42/21
omest, uppermost, higher, 35/26; omyst, 35/28
omwhile, sometimes, 45/31
on, one, 8/29
opyne, plain, 47/8
or, before, 13/25
or = þe oþer, the other, 28/34
ordure, order, 34/9; row, 43/1
other, or, 33/13, 43/26; other . . . or, either . . . or, 38/37
ouerer, upper, 42/15
ouer-hippede, passed over, 43/19

recte, directly, 27/20
remayner, remainder, 56/28
representithe, represented, 39/14
resteth, remains, 63/29
rewarde, regard, 48/6
rew, row, 4/8
rewle, row, 4/20, 7/12; rewele, 4/18; rewles, rules, 5/33

s. = scilicet, 3/8
sentens, meaning, 14/29
signifye(tyf), 5/13. The last three letters are added above the line, evidently because of the word 'significatyf' in l. 14. But the 'Solucio,' which contained the word, has been omitted.
sithen, since, 33/8
some, sum, result, 40/17, 32
sowne, pronounce, 6/29

singillatim, singly, 7/25
spices, species, kinds, 34/4
spyl, waste, 14/26
styde, stead, 18/20
subtrahe, subtract, 48/12; pp. subtrayd, 13/21
sythes, times, 21/16

taȝt, taught, 16/36
take, pp. taken; t. fro, starting from, 45/22
taward, toward, 23/34
thouȝt, though, 5/20
trebille, multiply by three, 49/26
twene, two, 8/11
þow, though, 25/15
þowȝt, thought; be þ., mentally, 28/4
þus = þis, this, 20/33

vny, unite, 45/10

wel, wilt, 14/31
wete, wit, 15/16; wyte, know, 8/38; pr. 2 s. wost, 12/38
wex, become, 50/18
where, whether, 29/12
wher-thurghe, whence, 49/15
worch, work, 8/19; wrich, 8/35; wyrch, 6/19; imp. s. worch, 15/9; pp. y-wroth, 13/24
write, written, 29/19; y-write, 16/1
wryrchynge = wyrchynge, working, 30/4
wt, with, 55/8

y-broth, brought, 21/18
ychon, each one, 29/10
ydo, done, added, 9/6
ylke, same, 5/12
y-lyech, alike, 22/23
y-myȝt, been able, 12/2
y-nowȝt, enough, 15/31; ynovȝt, 18/34
yove, given, 45/33
yt, that, 52/8
y-write, v. write.
y-wroth, v. worch.

The manufacturer's authorised representative in the EU for product
safety is Oxford University Press España S.A. of El Parque Empresarial
San Fernando de Henares, Avenida de Castilla, 2 - 28830 Madrid
(www.oup.es/en or product.safety@oup.com). OUP España S.A. also acts
as importer into Spain of products made by the manufacturer.
Printed and bound by CPI Group (UK) Ltd, Croydon, CR0 4YY

23/03/2026

02076308-0001